T0073217

SCIENCE AND BUDDHISM

Dialogues

SCIENCE AND BUDDHISM

Dialogues

Tsutomu (Jixin) Kambe

World Scientific

NEW JERSEY · LONDON · SINGAPORE · BEIJING · SHANGHAI · HONG KONG · TAIPEI · CHENNAI · TOKYO

Published by

World Scientific Publishing Co. Pte. Ltd.

5 Toh Tuck Link, Singapore 596224

USA office: 27 Warren Street, Suite 401-402, Hackensack, NJ 07601

UK office: 57 Shelton Street, Covent Garden, London WC2H 9HE

British Library Cataloguing-in-Publication Data
A catalogue record for this book is available from the British Library.

SCIENCE AND BUDDHISM
Dialogues

ISBN 978-981-125-851-0 (hardcover)
ISBN 978-981-125-852-7 (ebook for institutions)
ISBN 978-981-125-853-4 (ebook for individuals)

For any available supplementary material, please visit
https://www.worldscientific.com/worldscibooks/10.1142/12897#t=suppl

Typeset by Stallion Press
Email: enquiries@stallionpress.com

Contents

Prologue

The main themes of the present work are dialogues or the examination of whether there is a common platform between science and Buddhism. Both share their findings on this platform. This book is an attempt to verify the existence of such a platform. Let us begin with a story about a dialogue between a Buddhist sage and a king of the Indo-Greek kingdom in the BCE era.

Encounter between East and West

Gautama Buddha lived between the sixth and fifth centuries BCE (or c. 5th to 4th century BCE).[1] Greek communities developed in northern India one or two hundred years after the Buddha's time. This was the result of the expedition of Alexander the Great and his army. Bactria was an ancient Central Asian region that included modern-day Afghanistan. The Greek community developed into the Indo-Greek kingdom. The ancient Greek culture merged with the cultures of India, Western Asia, and Ptolemaic Egypt to form Hellenistic culture. In ancient India, the Indo-Greeks were known as "Yona," "Yonaka," or "Yavanas," which meant "Ionians." The Indo-Greek Kingdom flourished between the last two centuries BCE and the year 10 CE. It was before the ancient Pompeii was buried by the deadly volcanic current caused by the eruption of Mt. Vesuvius in 79 CE, so it was a time when Pompeii's people were enjoying their prosperity.

The famous ancient Buddhist sutra "The Questions of King Milinda" [1] dates back to around 100 BCE. This sutra has kept a record of many dialogues between the Buddhist sage Nāgasena and the Indo-Greek King Menander I (Milinda in Pāli) of Bactria in the 2nd century BCE. The King (probably when young) paid a visit to the sage Nāgasena and asked various questions about Buddhism. In the background of their dialogues and their location, there exist spectacular historical stories. The dialogues are

---

[1] According to the literature, it was 563–483 BCE (or 484–404 BCE).

xi

fascinating in and of themselves, but they are also very valuable from a modern point of view.

From the viewpoint of history, this was an encounter of two ancient influential philosophies: Indian Buddhist philosophy and Greek (Hellenistic) philosophy, the latter emphasizing logical reasoning. After more than two thousand years, our 21st century is witnessing a renewed encounter between modern science and developed Buddhism, both armed with the Wisdom of Emptiness and Great Compassion for all sentient beings. One of the main goals of this book is to document some interesting aspects of today's world through various dialogues between science and Buddhism.

Returning to the Buddhist sutra, a passage from the text states that King Milinda asked the sage Nāgasena: "Kindly explain to me as a Buddhist how human perception is characterized by grasping and how human wisdom is characterized by cutting off." The Buddhist sage responded to this question with an example:

> "Do you know how farmers reap barley? With the left hand they **grasp** the barley into a bunch, and taking a sickle into the right hand, they **cut it off** with the sickle."

This is an illustrative answer. The grasping here represents human perception of the surrounding world by one's sensory system, while the cutting off symbolizes Buddhist wisdom toward the affliction caused by sensing system perception and its handling into peace rather than chaos resulting from its direct response.

The people of the Indo-Greek kingdom were ardent recipients of Buddhism. In particular, the monk Dharmarakṣita [4] (in Sanskrit, protected by the Dharma) was one of the missionaries sent by the Mauryan Indian emperor Ashoka to spread Buddhism. In a chronicle, he is described as a Yona (not Indian), and his activities are indicative of some Hellenistic Greek during the early centuries of Buddhism. Dharmarakṣita became a respected elder Buddhist monk and settled in the Ashokan capital of Pataliputra, a port town along the Ganges. He then received a young Indian, Nāgasena, as a disciple and helped him toward awakening. Later, in his venerable age, Nāgasena met the young King Milinda.

The above explains the historical context in which the Buddhist text *Milinda Panha* was created. This book records the dialogues between a great Greek king (Menander I) and a Buddhist monk who was a disciple of the great Greek Buddhist elder Dharmarakṣita. This implies a certain important role of Greeks in Bactria during the first formative years of Buddhism.

Gandhara: Culture merging of East and West

Gandhāra was an ancient kingdom located within Bactria, in the Peshawar valley areas of what are now northwestern Pakistan and eastern Afghanistan. Gandhāra became independent from Bactria and formed the Indo-Greek Kingdom. King Menander I was the most famous. He became a Buddhist and is remembered in the Buddhist records mentioned above for his discussions with the Buddhist sage Nāgasena in the book *Milinda Panha*. Around the time of Menander's death in 140 BCE, the Central Asian Kushans overran Bactria and ended the Greek rule there.

Gandhāra's culture peaked during the reign of the great Kushan king *Kanishka* (127–150 CE). Purushapura became the capital of the great empire stretching from Central Asia to Northern India, with Gandhāra being in its midst. Emperor Kanishka was a great patron of Buddhists. Buddhist art spread from Gandhāra to other parts of Asia. Under Kanishka, Gandhāra became a holy land of Buddhism.

The Greco-Buddhist art or Gandhāra art of north India is the artistic manifestation of Greco-Buddhism, a cultural *merging* of Ancient Greek art and Buddhism. The series of interactions leading to Gandhāra art occurred over time, beginning with Alexander the Great and followed by the Mauryan Emperor Ashoka, who converted the region to Buddhism. Buddhism became prominent in the Indo-Greek kingdoms. Greco-Buddhist art truly flowered under the Kushan Empire, when the first surviving images of the Buddha were created during the 1st–3rd centuries CE. Figure 1 shows one such evidence. Gandhāra art reached its zenith between the 3rd and the 5th century CE (*Greco-Buddhist art*, Wikipedia). During the earliest period of Buddhism, when the Buddha was not represented with statues like Fig. 1, the Buddha's footprint (Fig. 2) was one of the symbols used in narrative reliefs depicting the Buddha's life scenes.

Fig. 1   *Tathagata Statue*, Indo-Greek style, in Gandhāra (1st –2nd CE). Photograph by H. Kikuchi [87] at Tokyo National Museum (2017).

Fig. 2   Buddha footprint (*Buddhapada*): On the path from aniconic to iconic, which starts at symbols like the wheel and moves to statues of Buddha, the footprints are meant to remind us that Buddha was present on earth and left a spiritual path to be followed. Courtesy of Ven. Sangharatna Manake (Pannya Metta Sangha, Nagpur, India).

Buddhists as human beings

Now let us consider another important topic. The main concern of Buddhism is reality (which is *timewise*, not at a single moment), i.e. faithful realization of our interrelationship with the world surrounding us, while science is concerned with the dynamical reality and diversity of nature that can be tested empirically. To approach the reality of the world, both Buddhists and scientists share the same stance in the sense that both do not rely on miracles or mysterious events that are thought to have been caused by an unidentified or supernatural power. This is said to be *scientific*. In fact, a theory is said to be *scientific* if it can be empirically verified. Experimental verification is an essential component of modern science.

Buddhists see themselves as an integrated system that is conditioned by interaction with their surrounding world. They refine themselves constantly through physical exercise and mental activity, and they keep refreshing themselves ceaselessly through deep meditation and *samādhi*. Scientists usually rely on empirical evidence that can be tested using specially designed apparatus for *experimental verification*, while Buddhists verify themselves by relying on their own living bodies and staying mindfully in the awakened state.

Observing ourselves and our daily life every moment, we experience the continuous change of both ourselves and the external phenomena around us. The external world is changing its appearance ceaselessly, as is our perception of it through our sense organs. Both (*internal* and *external*) are arising and ceasing one after the other. Similarly, our minds are continuously due to the emergence and cessation of thoughts. All of these originate depending on each other. All phenomena, whether *internal* or *external*, are viewed as *streams*. In the Buddhist view of world phenomena, the streams are captured as consisting of twelve factors and are called the *twelve links of interdependent origination* (*twelve nidāna*s) ([2, 3], *Abhidharmakośa-bhāṣya* of Vasubandhu, Chap. III, p. 1030])

According to the Buddhist philosophy, it is understood that all of these give rise to two types of defilements: *desires* and (off-point) *false* views. A defilement itself is, however, an active realistic state of living minds, characterized by Buddhistic terms such as *ignorance, craving*, and *grasping*. Its nature is explained to be like that of a seed, a root, or a tree.

From the seed, there arises a sprout, stalk, leaves, etc. In the same way, from defilement there arises another defilement, action, and phenomenal basis. A tree continues to grow if its root is not cut off even though it is pruned again and again. In the same way, as long as the defilement is not cut off, the state of mind continues to grow. A tree grows to produce flowers and fruits at different times. In the same way, a defilement grows to produce another defiled action and to ripen the fruit of the defilement. The streams of defiled existences of human beings are rooted in fundamental ignorance (*avidyā*).

[Comment: "*Vidyā*" in Sanskrit means "science" or "valid knowledge." Hence "*avidyā* means "unknown scientifically.""]

It is remarkable, according to the above Buddhist view, that cutting off the defilement is within their sight — in other words, Buddhists regard it as an important subject. According to orthodox Buddhist philosophy, the streams of world phenomena in the human world are captured as the "twelve links of interdependent origination," which are as follows:

Ignorance (nescience, *avidyā*) → **Karma formation** → Consciousness (*vijñāna*) → Name-and-form → Six sense-bases (*eye, ear, nose, tongue, body,* and *mind*) → *Contact* → *Sensation* → *Craving* → *Grasping* → **Existence** → Birth → Aging and death.

The intrinsic nature of the above twelve factors' dependent origination is threefold: *defilement, action* (karma), and *basis of phenomena*: (i) Three members (i.e. ignorance, craving, and grasping) are intrinsically defiled. (ii) Two factors (i.e. karma formation and existence) represent *action*. (iii) Seven factors (i.e. consciousness, name-and-form, six sense-bases, contact, sensation, birth, and aging and death) are intrinsically the *basis of phenomena* and thus qualified because they are the source of defilements and actions. The factors of a phenomenal basis are effects or bases, while the factors of defilements and actions are causes in their intrinsic nature.

At the beginning of this Prologue, a passage from a Buddhist text is cited, in which the sage *Nāgasena* says, "Human wisdom is characterized by cutting off." In addition, in another Buddhist text, Master Vasubandhu describes a Buddhist's view on the defilements being associated with humans' perception of their surrounding world. Buddhists have a philosophical theme of "cutting off" their own defilements because extinguishing the fire of defilements is essential for attaining the Supreme Way. Hence, Buddhists are encouraged to have the wisdom of "cutting off" within sight.

Gautama's renouncement is considered as a kind of the Buddhist wisdom of "cutting off" considered above. It is not a simple "cutting off", but a prerequisite for later awakening, which would have not been accomplished without the "cutting off". In other words, Gautama Siddhārtha attained awakening through his *śramaṇa* life after renouncement of former household life and became *Gautama Buddha*.

Shakyamuni Gautama Buddha

The Shakyamuni Gautama Buddha renounced his princely life at the age of 29. Gautama Siddhārtha decided to leave his palace in order to live the *śramaṇa* life of a wandering ascetic. Before renouncing, he asked his father, King Suddhodana of the Shakya clan (1st millennium BCE), to permit him to leave the palace and become a *śramaṇa* monk to seek the way of utmost liberation from suffering. On a bright, moonlit night, he quietly left the palace on the back of his horse Kanthaka, accompanied by Channa on his own horse up to a halfway point. Gautama Siddhārtha told himself, "If I do not find the Way, I will not return to Kapilavatthu (his hometown) again".

Having arrived at the edge of a forest after crossing a river, Siddhārtha dismounted from his horse Kanthaka. Stroking Kanthaka's mane, "Siddhārtha pulled out a sword tucked in his horse's saddle and then grasping his own locks of hair in his left hand, cut them off with his right." Siddhārtha handed his attendant Channa the hair and sword, and told him, "No monk has need of a personal attendant. Please, return home now"!" ([6]: Thich Nhat Hanh (1991), *Old Path White Clouds*).

At the age of 35, beneath a pippala tree (bodhi tree, located in Bodhgaya, Fig. 3), Gautama sat down in a lotus position and entered into deep dhyāna meditation of great concentration overnight. Looking up, Gautama saw the morning star shining on the eastern horizon. This is the moment of Great Awakening.

Fig. 3   The historic pippala tree in Bodhgaya. This large pippala tree (pipal tree, bodhi tree) stood at the western side of the Mahabodhi temple in Bodhgaya. Gautama Siddhartha Shakyamuni sat underneath the pippala and entered into deep dhyāna meditation some 2500 years ago (https://bodhitree.in/).

Later, at the age of about forty, Gautama Buddha returned to his hometown, Kapilavatthu, and met his father, King Suddhodana, Queen Gotami, his former spouse Yasodhara, and their son Rāhula. Rāhula was a newborn baby when Siddhārtha had left his palace. Probably, Siddhārtha had been reminded of his real mother, former Queen Mahamaya, who had passed away while giving birth to Siddhārtha himself — a profound sorrow he could never forget.

Queen Gotami was a sister of Mahamaya. Later, when King Suddhodana passed away, Gotami decided to renounce. After another long story, Gotami succeeded in being ordained by Gautama Buddha and became the first bhikkhuni (nun), Mahāpajāpatī, along with hundreds of lady attendants, including Yasodhara. It was some time after that her son Rāhula became a novice monk.

Bodhidharma

In regard to the symbolic "cutting off" of Buddhists, another amazing historic event would be worth mentioning. It happened a thousand years after the *parinirvana* of Gautama Buddha Shakyamuni. According to Chinese literature [54, 55], Bodhidharma (c. 440–536 CE), the founder of Zen Mahāyāna Buddhism, was born as the third prince of a kingdom in South India that was recorded only by three Chinese characters as 香至国. Now, it is understood to denote the ancient Indian ancient town Kancheepuram (see Part III), where the capital of the Pallava dynasty was located in the area of what is Tamil Nadu State today. As a child, he was called Bodhidhara.

Bodhidhara was a person of great dharma talent. When his father, a king, passed away, Bodhidhara renounced his life as a prince (perhaps before he turned 20) and was given the Buddhist name Bodhidharma by his dharma teacher Reverend Prajnadhara, the most respected monk in those days and the twenty-seventh Patriarch in the authentic lineage descending from the first master, Maha-Kassapa, inheriting the dharma of Shakyamuni Buddha himself.

According to Chinese literature [54], the Reverend Prajnadhara entered *parinirvana* in the year 457 CE. Now, Bodhidharma was the twenty-eighth Patriarch succeeding the authentic Dharma of Shakyamuni Buddha. Thereafter, Bodhidharma traveled throughout India and turned the Buddha Dharma Wheel for her people. During his time as an itinerant monk, Bodhidharma had never forgotten his teacher's advice to visit China for authentic Mahāyāna Buddhism. Much later, more than sixty years since the *nirvāna* of his adviser-teacher, he realized that the time had come for his visit to the East-land (China) as a monk of Mahāyāna Buddhism.

A Chinese text [59] describes what happened when he determined to leave India. First, he bade farewell at the stupa of Ven. Prajnadhara. Then he met the King and told him of his wish to go to China. The King persuaded him to stay in his home land, but Bodhidharma was determined to go. There was no longer anything for the King to do but to prepare a big boat with necessary goods for a safe voyage. He hoped that Bodhidharma would return in the future (preferably with the same ship). On the day of departure, the King accompanied Bodhidharma up to the harbor (Mamallapuram probably), along with the families of his relatives and high-ranking officials. At this moment, there was no one who was not in tears [54].

This was the moment that might be said to be a typical Buddhist *cutting off*. In the case of Bodhidharma, since he had already renounced the palace life, his leaving the home country could be interpreted as the Buddhist *cutting off* for future development.

Although it is not widely known, during his stay in China, Bodhidharma left written texts for his disciples, which were translated into Chinese by his disciple Huike. Only the Chinese version [75] is now available for us (see Part III of this book). According to it, it took him three years to travel from India to China by sea. In the author's view, spreading Buddhism in China requires finding the best disciples for Bodhidharma deliberately, although his aim was implicit (but apparent).

What happened around Bodhidharma was a sequence of amazing dramas, comprising fourfold upholding stories of *cutting off*, which culminated in the explosive development of Chan (Zen) Buddhism in Chinese society. The first two have already been presented: (*i*) renouncing his life as a prince and (*ii*) leaving his home land, India. The third was a story concerning how he could acquire the best Chinese disciple at Shaolin.

In Luo-yang (an ancient capital city located south of the Yellow River), there was a prominent scholar named Shen-guang (神光 later HuiKe Da-shi) who was learned in all the teachings of Con-zi, Lao-zi, and Zhang-zi. Hearing that a monk of high virtue, Da-shi (大師 great teacher in Chinese), who had come from India, was staying at the Shao-lin monastery, Shen-guang came to pay homage to the monk in order to hear and to master the Supreme Way of Shakyamuni. He reverently requested Da-shi Bodhidharma to teach him the Buddha Dharma. However, whenever he tried to ask questions morning and evening, he received only silence. One day in December, as snow was falling, Shen-guang remained standing all night in the garden in front of the place where Da-shi was sitting (Fig. 10.5). The snow lay knee-deep. At last, after midnight, the teacher (Da-shi) opened his mouth:

"Why you have been standing in the snow?"

Fig. 4 (a) The highland Kamuy Mintar (Hokkaido, Japan) and the *komakusa* flower (inset): https://icotto.jp/presses/8859 (b) *Usubakicho* (butterfly) perching on a *komakusa* (flower) [8].

Shen-guang replied almost in tears: "Da-shi, please open the Dharma Gate with great mercy and compassion, and save all sentient beings globally."

Suddenly, Shen-guang took up a knife and *cut off* his left arm in front of the teacher, leaving him standing in the snow. The literature [56] records what Shen-guang later told his disciple about this snowy night:

"When I was practicing to seek the Buddha Dharma here, I cut off one of my arms and stood in the snow from first-night [about eight p.m.] to third-night [midnight], and did not recognize that snow became over my knee. In this way, I quested for the Supreme Way."

It is said that the snow around him was dyed red at the time. No doubt, Bodhidharma was deeply moved by this action of his disciple, and it impressed him with the depth of the disciple's desire to seek the Supreme Way.

Master Bodhidharma, the Great Teacher, had kept his disciple's precious wish in his mind throughout his entire later life. It appears that this was made explicit at the last moment of his life. On the basis of Chinese literature, a detailed account would be required to describe the story precisely, and hence it is left to the main text (Section 9.3.8). In fact, this is the fourth story of *cutting off*.

So far, we have had an overview about *how human wisdom is characterized by cutting off*, with dramatic stories of the theme *cutting off* in human societies.

Natural world

Next, let us see whether one can find any wisdom in the natural world by looking into the world of the flower-and-insect, as an example. At the high altitude 2,000 m in Japan, the coexistence of lovely alpine flowers and alpine butterflies can be observed at an alpine natural tapestry of the Daisetsu-zan Mountain Range (once a big volcano) located at the center of Japan's northern island Hokkaido. In their language, the local Ainu people there call the range Kamuy Mintar (registered as a Japan heritage site), meaning a "playground of gods and spirits" [7]. They have been the original inhabitants of northern Japan for more than ten thousand years, living a "hunting-gathering" life and regarding their collected foods as gifts from gods of mountains, gods of trees, gods of rivers, or gods of volcanic fires. They also regard wild bears as a blessing of nature, and when presented with a bear, they express their gratitude by dancing.

The mountain range Kamuy Mintar is covered with snow for nine months of the year, and there are only two seasons there: winter and summer. Once the snow has melted and disappeared, alpine flowers and insects wake up to work actively and ceaselessly during the short summer.

An amazing story of the coexistence of the lovely alpine flower *komakusa* (*Dicentra peregrina*, a herbaceous plant) and alpine butterfly *usubakicho* (a mountain Apollo butterfly of the swallowtail family) was observed and recorded. In summer, the alpine butterfly *usubakicho* lays eggs on the leaves of the alpine *komakusa* plant. However, the egg state remains for about 11 months; hence the eggs spend the entire winter in the snow before hatching the following summer. Then the egg becomes a larva. What is more, the larva eats the same plant exclusively. In fact, both the alpine flower plant and alpine butterfly are survivors of the glacial period. After the glacier receded to the polar region, they could find their ecological niche in the highlands.

In general, the butterfly undergoes a complete metamorphosis, from an egg passing through a larval stage, then entering an inactive *chrysalis* pupal stage, and finally emerging as an adult butterfly. This is usually completed within a year. However, the case of the alpine butterflies is different: they remain dormant as the egg stage in the first winter and thereby survive the severe snowy winter. This is their wisdom, isn't it? Because if their eggs become larvae before the coming winter, it would be almost impossible for the larvae to survive until the next summer. The larval period of the Apollo butterfly is three months. The larva becomes a pupa during the second summer time and then spends the second winter for about ten months. In the third summer, the alpine butterfly finally emerges as an adult.

What is the merit of the unique coexistence of the lovely alpine flower *komakusa* (a herbaceous plant) and the alpine butterfly *usubakicho* (a mountain Apollo butterfly)? The Apollo butterfly is known to produce a repulsive taste for its predator. The butterfly seems to get this foul taste from its plant host. In addition, a recent web article reports it as follows. Two Japanese pharmacologists (one in 1909 and the other in 2003) collaboratively found that toxic ingredients (morphine-like substance) are included in the alpine plant *komakusa* [8]. Hence, the larvae feeding on the plant leaves are likely to produce a repulsive taste for its eaters or plant-eating animals. Thus, the alpine plant is protected from plant eaters. This is their strategy and the secret of coexistence.

The unique coexistence of the alpine flower *komakusa* and the alpine butterfly *usubakicho* was highlighted in 1985 by a TV documentary of NHK Japan on the natural wildlife [9]. Its narrator began: "Can you believe the fact that flowers have wisdom? They really have 'wisdom.' " The TV presented a number of pictures of beautiful scenes showing wild life at Kamuy Mintar: flowers, insects, birds, a rock rabbit squeaking, and natural surroundings, which were captured by cameramen who stayed there for the short summer of two months. Finally, the narrator stated, "I have been convinced that they have wisdom." In fact, each species has its own wisdom for surviving the long, severe, snowy winter.

This reminds us of the Buddhist tradition of the *rain retreat* (Vassa tradition), which takes place during the wet season of India lasting for three months. Learning from the natural world might be said to be Buddhist wisdom. The rain retreat was, however, a long-standing custom for mendicant ascetics in India predating the Buddhist times. The retreat is a kind of cutting off of the continuously connected sequence of lifestyle,

Fig. 5 Bodhidharma Memorial. This was erected in July 2019 in his hometown, Kancheepuram, the capital of the ancient Pallava Kingdom, located in South India.

which is in fact wisdom, as seen above, that prepares a creative state for the subsequent future.

In this publication, various forms of mutual dialogue between science and Buddhism or between different human societies or different cultures are presented mindfully. In this way, scientific aspects of humanity are illuminated in the light of modern science, Buddhism, and various philosophies. Modern science is reviewed in the light of Buddhism,

Comemoration Project of Bodhidharma

The second theme is the Tamil muni Bodhidharma, a Buddhist monk from the 5th to 6th century CE. He crossed the oceans from his home land, India, traveling as far as China while firmly holding the Mahāyāna spirit. He founded Zen Buddhism in China, which is nothing different from the authentic Buddhism of Buddha Shakyamuni. If any difference is required to mention, it could be the living styles of the sages in various human societies.

To commemorate his unsurpassed achievement, a Bodhidharma Memorial Stone Stele was erected in 2019 in his hometown, Kancheepuram (India). This project was carried out under the leadership of a non-profit organization (registered in Japan) in collaboration with local supporters in India. The author, a person engaged in the project, reports the project from its initiation. Chapter 9 and the Part III describe not only the commemoration project but also historical stories from Chinese literature about illuminating dialogues between Bodhidharma and Chinese Buddhists. The Bodhidharma's teachings recorded in Chinese by his disciple monks are very precious. Some of them are included in this publication.

Author's words

In this book, the author provides a bird's-eye view of his lifelong study of science and Buddhism as a physicist and Zen Buddhist since his youth. In other words, scientific aspects of humanity and Buddhism are illuminated in the light of modern science and physiology, while modern sciences are reviewed in the light of Buddhism. Various forms of mutual dialogues between science and Buddhism, between societies of different cultures, or within themselves, are mindfully presented in this publication.

Buddhism and Science: Dialogues

PART I

Wisdom of Shakyamuni Buddha

Chapter 1

Ancient Dialogues of Two Wise Men
Examination of names and self

The *Milindapañha*

Dialogues between King Milinda and the Buddhist sage Nāgasena

Fig. 1.1. Fault text edited by V. Trenckner, a Danish pioneer in Pāli lexicography. Williams and Norgate, London, 1880.

Fig. 1.2   King Milinda is posing subtle and knotty questions to the Buddhist sage Nāgasena, wisely turning on many points. [***Milindapañha***, 140 BCE] (drawn by a Vietnamese Buddhist Monk, father of Nun Nguyen Thi Ngo, 2016).

Starting the first chapter of *Buddhism and Science: Dialogues*, it would be most appropriate to begin with the historic dialogue exchanged between two distinguished persons, by citing the sutra *Questions of Milinda* (*Milindapañha*), because this text records the ideal *encounter* of the Buddhist sage Nāgasena and King Menander I (Milinda in Pāli text) of the ancient Indo-Greek nation Bactria in the 2nd century BCE. The King (young, probably) visited Venerable Nāgasena and asked various critical and valuable questions on Buddhist philosophy (Fig. 1.2). There were spectacular stories concerning how this encounter happened in regard to the historical age and geographical location. See Prologue for some of these stories.

Given that the time was the BC ages, the dialogue was just a wonderful record in human history. However, our concern is how the dialogues were carried out and how the two philosophies of quite different backgrounds were communicated to each other by two wise men. It will be quite interesting to know whether the philosophical idea was conveyed successfully between them.

In the 2nd century BCE, the Indo-Greek King Menander I (Milinda in Pāli) of Bactria had a strong interest in Buddhism, and his intellectual dialogues with the Buddhist sage Nāgasena were preserved in a famous classic sutra, the *Milindapañha*, *"Questions of Milinda"* (Menander). It was after the time of Asoka. Shortly after the death of Asoka the Great in 232 BCE, the Maurya Empire broke up and the north-western frontiers of India beyond the Indus (Bactria and Parthia) came to be held by rulers of Greek descent. King Menander of Kabul penetrated far into Northern India and established a capital at Sagala (Sialkot in the Punjab).

Milindapañha (Questions of Milinda)

Opening *narration*

King Milinda, the King at Sâgala of the capital town of Yonaka
(Yona, Ionean, Greek),
has visited the Monastery, where Nâgasena is staying, the renowned
sage.
So as if the Ganges flows out into the Deeper Ocean.
The King who is known to be eloquent, to Nâgasena who is a bearer of
torch of Truth with calmness of minds, did put subtle and knotty questions
concerning truth and falsehood, and turn on wisely many points. Then,
Sage's answers to the questions were based on profound philosophy, sweet
to hearers' ear and pleasing the heart, and were something that has never
been done before, Passing Wonderful and Strange.

Nâgasena's talk plunged to the hidden depths of Vinaya (rules) and of
Abhidhamma (Dharma-philosophy, Chap. 2), unravelling all the meshes
of the Sutras' net, glittering the while with metaphors and reasoning high.
Come then! Apply your wisdom, and let your hearts rejoice, and hearken
to these subtle questionings, such that all grounds of doubt would be
resolved. [1, 10]

1.1 *Milindapañha*: (Chap. 1) Individuality and Name

According to the references [1, 10, 11], chapter 1 of the sutra begins as follows:

> King Milinda went up to where Venerable Nāgasena was, and exchanged greetings politely and friendly to each other and took his seat respectfully to one side.

King Milinda began by asking the Venerable Nâgasena about the fames and names of the Monk Nāgasena:

> *"How is Your Reverence known, and what is your name, Sir?"*

The Venerable *Nâgasena* said to the King:

> *"I am known as Nâgasena, and it is by that name that my brethren in the faith address me, 'Nâgasena'. But parents give such a name as Nâgasena, or Sûrasena, or Vîrasena, or Sîhasena, yet this is only a naming designation in common use. In fact, it's just, a common name, a provisional name, a tentative name or a given name. No real entity is found there, because there is no permanent individuality (No Soul) underlying there. Human being is destined to die and its body disperses like mist or cloud."*

Then Milinda called upon Yonakas (his attendants) and the brethren (of the sage Nāgasena) to witness:

> *"This Nâgasena says there is no permanent individuality (no soul) implied in his name. Is it now even possible to approve him in that?"*

Fig. 1.3   An illustration from [11].

Turning to Nāgasena, he said:

"If, *Reverend Nâgasena*, there be no permanent individuality (no soul) involved in the matter, who is it, pray, who gives to you (a member of the order) the four necessaries: robes, foods, bed and sitting mat, and medicines for illness? Who is it who enjoys such things when given?

"Who is it who lives a life of righteousness? Who is it who devotes himself to meditation? Who is it who practices the excellent Way and attains the Nirvâna of Arahatship?

"And who is it who destroys living creatures? Who is it who takes what is not his own? who is it who lives an evil life of worldly lusts, who speaks lies, who drinks liquors, who commits one of five sins leading to hell of uninterrupted suffering (*Avici* hell).

"If that be so (because of no personal individuality), there is neither merit nor demerit; there is neither do-er nor causer of good or evil deeds; there is neither fruit nor effect of good or evil Karma.

"*Most reverend Nâgasena*, if we are to think that, were a man to kill you, there would be no murder, then it follows that there are no real masters or teachers in your Order, and that your ordinations are void.

"You tell me that your brethren in the Order are in the habit of addressing you as Nâgasena. Now what is that Nâgasena? Do you mean to say that the hair is Nâgasena?"

"*I don't say that, Great King.*"

"*Or the hairs on the body, perhaps?*"

"*Certainly not, Great King.*"

"*Or is it the nails, the teeth, the skin, the flesh, the nerves, the bones, the marrow, the kidneys, the heart, the liver, the abdomen, the spleen, the lungs, the large intestines, the small intestines, the stomach, the feces, the bile, the phlegm, the pus, the blood, the sweat, the fat, the tears, the serum, the saliva, the mucus, the oil that lubricates the joints, the urine, or the brain, or any or all of these, that is Nâgasena?*"

And to each of these, he answered *No*.

"Is it then the body-form (*Rûpa*) that is Nâgasena, or the sensations (*Vedanâ*), or the conception (*Saṃjñā*), or the mental formation or volition (*Saṃskāra*), or consciousness (*Vijñāna*), that is Nâgasena?"

And to each of these also, he answered *No*.

"Then is it all the five Skandhas (*form, sensations, perceptions, mental formations,* and *consciousness* [see Section 2.5]) combined that are Nâgasena?"

"No! Great King."

"But is there anything outside the five Skandhas that is Nâgasena?"

And still he answered *No*.

"Then thus, though I have asked you a number of times, I can discover no Nâgasena. Nâgasena is a mere empty sound. Who then is the Nâgasena that we see before us? It is a falsehood that your reverence has spoken an untruth by answering *No*."

Venerable Nāgasena said to the King, asking questions in return.

"You have been brought up in great luxury, as beseems your noble birth. If you were to walk this dry weather on the hot and sandy ground, trampling under foot the gritty, gravelly grains of the hard sand, your feet would hurt you. And as your body would be in pain, your mind would be disturbed, and you would experience a sense of bodily suffering. How then did you come, on foot, or in a chariot?"

To these questions of Venerable Nāgasena, the King replied:

"I did not come, Sir, on foot. I came in a carriage."
"Then if you came, Sir, in a carriage, explain to me what that is. Is it the shaft that is the chariot."
"I did not say that, Sir."
"Is it the axle that is the chariot?"
"I did not say that, Sir."
"Is it the wheels, or the framework, or the ropes, or the yoke, or the spokes of the wheels, or the whip, that are the chariot."

And to all these, he still answered *No*.

"Then is it all these parts of it that are the chariot?"
"No, Sir."
"But is there anything outside them that is the chariot?"

And still he answered *No*.

"Then Sir, though I have asked you a number of times, I can discover no chariot. Chariot is a mere empty sound. What then is the chariot you say you came in? There is no such thing as a chariot! It is a falsehood that your Majesty has spoken, an untruth! You are king over all India, a mighty monarch. Of whom then are you afraid that you speak untruth?"

Then, Venerable Nāgasena called upon the Yonakas (five hundred Greek attendants) and the brethren to witness, saying:

"King Milinda here has said that he came by carriage. But, when asked in that case to explain, what the carriage was, he is unable to establish what he averred. Is it possible to approve him in that?"

When he had thus spoken, the Yonakas shouted their applause, and said to the King:

"Now your Majesty, please get out of that if you can?"

King Milinda replied to Nāgasena and said:

"I have spoken no un-truth, reverend Sir. It is because it has all these parts (you questioned me about) — the shaft, the axle, the wheels, the spokes, and the whip — that it comes under the term *'chariot'*. The 'chariot' is a generally understood term and *the designation in common use*."

"Your Majesty, very good! You have rightly grasped the meaning of '*chariot*'. And just so, it is because of all those things (you questioned me about) — the thirty-two kinds of organic matter in a human body, and the five Skandhas constituting the being — that I come under the name 'Nâgasena'. The name is a generally understood term and *the designation in common use*. For it was said, Sir, by our Sister *Vagira* in the presence of *Buddha* the Blessed One:

> "*Just as it is by the condition precedent of co-existence*
> *of its various parts that the word 'chariot' is used,*
> *just so, is it that when the five Skandhas are there we talk of a 'being'*

"*Most wonderful, Nâgasena, and most **strange**, Well has the **puzzle** put to you, most difficult though it was, been solved. Were the Buddha himself here, he would **approve your answer**. Well done, well done, Nâgasena!*"

1.2 "Strange" or "Excellent"?

The dialogue between the two wise men finally reached a complete understanding. However, the translators show a subtle difference between the English translation [1] and the Japanese translation [10]. As regards the last statement of King Milinda, the English version is like the last four lines, while the Japanese version, if translated into English, is as follows:

> "*Most wonderful, Reverend Nâgasena. It's **excellent**, Reverend Nâgasena. The **question** put to you has been resolved very well. Were the Buddha himself here, Buddha would **give you a word of praise**. Well done, well done, Nâgasena! **The question has been resolved very well.***"

As can be seen from the boldface words in both expressions, one recognizes **strange** and **puzzle** in the former version, while in the latter version the corresponding parts are **excellent** and **question**. Although the latter version expresses complete approval, the former gives the readers an impression that the speaker's response is reluctant. In addition, the Buddha's response is simply "***would approve your answer***" in the first version, while the latter version not only writes "***would give you a word*** *of praise*" but repeats the statement "***The question has been resolved very well***" at the end. These differences are observed in the translations from the same original text.

The main reason for the difference is probably that the author(s) of the sutra *Milindapañha* was (were) definitely Buddhist(s), and that its translation depends on whether Buddhist philosophy is widely accepted by the society of the translator or not. We must realize that this kind of difference always occurs in the dialogues exchanged between people in the Eastern world and Western world.

Since the difference might be caused by Buddhist philosophy, plausible parts of Buddhist philosophy are presented in the next section in order to clarify the details.

1.3 Buddhist Philosophy: *Dependent Origination and Empty of Self*

The central concepts of Buddhist philosophy are as follows: (i) What is happening in the world is characterized by the concept of *dependent origination*. This states that everything arises depending upon multiple causes and conditions and that nothing exists as an independent entity. As an example of dependent origination, let us consider a pot.

A pot is made from clay under the supporting conditions of water and fire heat. A pot has no *entity* in the sense of an inherent existence or self-existence. Why? According to conventional reasoning, a pot is *effected* from the *cause* of clay by the *scientific conditions* of water and heat. It does not make sense to consider that the pot (as an effect) pre-existed in the material of clay (as a cause). Moreover, before the pot is made, it does not make sense to consider that clay and water are causal conditions for the pot, because those would be conditions for non-existence (because the pot is not there). In addition, if the pot is heated to a temperature much higher than that at which it was made, it will melt into a liquid or, at even higher temperatures, will become gas and eventually diffuse into space. This implies that the pot has no *entity* and is not thought to have an inherent existence or self-existence.

Concerning **empty of self**: Around 150 BCE, Buddhists had a debate about how the ātman *(self-soul)*, or *pudgala*, *is denied as a real entity*. The Sanskrit words *pudgala* denotes a person, personal entity, ego, or soul.

There is visual consciousness. Conditioned by both the eye and the visible object, visual consciousness arises. This visual consciousness can only perceive what is visible, not the *pudgala* (the self-soul). Compare this with the fact that the triplet of the visual contact (*eye, visible object,* and *visual consciousness*) corresponds to ②, ①, and ④ in Fig. 3.2 (Chapter 3).

A *pudgala* cannot be perceived by visual consciousness, which is nothing but the transmission of neural signals excited by the incoming light rays; only what is visible is perceived by visual consciousness. Thus, this visual consciousness is not the *pudgala*. Contact is made by the coming together of three elements: the eye, the visible one, and visual consciousness. Conditioned by contact, there is sensation. This sensation generated by visual contact can only sense what is visible, not the *pudgala*. The *pudgala* is not sensed by the sensation generated by visual contact; only what is visible is sensed by the sensation generated by visual contact. Thus, this sensation generated by visual contact is not a sensation by the *pudgala* ([2], Vol. I (p. 58) by Bhikkhu Dhammajoti).

The observation of ancient Buddhists was certainly scientific from a modern point of view. To summarize the above argument:

> *Hence, the pudgala does not exist. As with the visual consciousness, the same is true for other five cases of auditory, olfactory, gustatory, bodily and mental consciousnesses.*

There was a school asserting that, while there is no ātman (self-soul), there exists a *pudgala* (person) or *sattva* (being). It was in the 4th century CE when Vasubandhu wrote his objection to the notion of *pudgala* (a person, personal entity, or ego) in his masterwork [2], by saying to their theory of *persons* that *they see a self in what is not a self and fall into the abyss of views.*

Vasubandhu writes: "Buddha explained that what we call a person (pudgala) is simply the aggregates," which arise depending on conditions coming together, and cites a passage from the text [2] as follows:

> *The self, in fact, does not have the nature of a self. It is through a mistaken mental construction that one thinks that a self exists; there exists*

no self (ātman), but only the dharmas produced by causes such as the five aggregates [Chapter 2] or the twelve members of dependent origination [Chapter 4]. When examined in depth, no person (pudgala) is perceived in them. See internal phenomena, which are empty of self; See external phenomena, which are empty of self, and even the practitioners who meditate on emptiness do not exist as an real entity.

[2, Chap. IX, p. 2538]

Chapter 2

Ancient Scientific Philosophy of Buddhists: *Abhidharma*

Master *Vasubandhu*
(c. 350 – 430 CE)
Kōfuku-ji temple,
Nara, Japan

2.1 Introduction: What is *Abhidharma*?

In the ancient times of BCE and early CE (common era) centuries, Indian Buddhists had scientific views (philosophy) concerning the world they were living in, and also about themselves. Their views were collected as the *abhidharma* theory. However, the whole documentation is too massive for a single person to capture the entire picture of the philosophy. The *abhidharma* theory is well known among most Buddhists. However, it

15

would be fair to say that only a few can provide the entire picture of the philosophy. One of the reasons may be that there are a huge number of arguments in the theory and that some are reliable while others are not, from modern point of views. The Sanskrit original of *abhidharma* theory was lost for centuries and was known to scholars only through Chinese and Tibetan translations. However, the philosophy was very important to the history of Indian thought.

First, the present author would like to express deep gratitude to Venerable Gelong Lodro Sangpo, who recently translated the masterpiece *Abhidharmakośa-bhāṣya* of Vasubandhu into English, and also to the publisher Motilal Banarsidass. The *Abhidharmakośa-bhāṣya* was a great work of master Vasubandhu (c. 350–430 CE), written 1600 years ago. In recent modern age, it was translated first into French by the great scholar Louis de La Vallee Poussin (1869–1938) and then into English (2012). Without the precious work done in its 2012 edition [2], it is certain that Chapters 2, 4, 5, and 6 of the present book would not have come into being.

To start Chapter 2, let us first ask the following question: What is meant by the term *abhidharma*? In Sanskrit, the prefix *abhi* signifies *abhimukha*, meaning "in the presence of" or "to the direction of." Hence, as a whole, *abhidharma* signifies "direct realization of dharmas" or "putting into the true nature of dharmas."

By definition, *abhidharma* means pure *prajñā*, i.e. realization of wisdom without defilement. Both *abhidharma* and *prajñā* mean realization, investigation, or discernment of dharmas. Here, dharma does not mean "doctrine," but refers to either "the ultimate reality" or "*nirvāṇa* as the dharma par excellence," or sometimes "phenomena" in general. Dharma is called *abhidharma* because it induces all dharmas conducing to "enlightenment" directly and effectively [3]. *Abhidharma* is stainless understanding, coexisting with our five aggregates ("aggregate" to be defined in Section 2.3). Therefore, it is real-world understanding. The *kośa* of *Abhidharmakośa-bhāṣya* means that this treatise is like the treasury (a treasure house) of *abhidharma*, and *bhāṣya* means a commentary.

2.2 Scientific Approach to the World Phenomena

In Indian tradition, learned persons are supposed to be well versed in five sciences. Such a scholar is given the title *paṇḍita*. The great master

Vasubandhu was one of the *paṇḍita*s. He compiled the massive corpus of *abhidharma* into *Abhidharmakośa* (Treasury of *Abhidharma*) [2, 3], according to the following view:

> *Without full discernment of dharmas, there is no method to quell the afflictions. Because of the afflictions, the worldly beings accumulate karma and wander in the ocean of existence, and do not achieve freedom (liberation).*

Four of the five sciences are linguistics, logic, medicine, and arts and crafts, and the fifth is the inner science that leads one to the ultimate realization of the self. Because of the fifth science, the ancient Indian philosophy is unique. Usually, wise individuals practice *Tripiṭaka*, the three "baskets" of Buddhism: *sūtra*s, *vinaya* (monastic rules), and *abhidharma* (given at the beginning). A Mahāyāna sūtra states:

> *Without becoming a scholar in the five sciences, not even the supreme sage can become omniscient. For the sake of refuting or supporting others and in order to know everything of oneself, one makes an effort in these subjects.*

Both *Abhidharmakośa* and science share the same stance to approach the reality of the world, in the sense that both do not rely on miracles or mysterious events that are thought to have been caused by a supernatural power or an unidentified one. This approach might be considered scientific. Master Vasubandhu, who was a prominent Buddhist and one of the *paṇḍita*s, is the author of the treatise *Abhidharmakośa*, and *Abhidharmakośabhāṣya* is his auto-commentary on the work.

In this chapter, the first two sections provide an introduction of the author, Master Vasubandhu, and describe the definition of the six sense organs. The subsequent sections give twofold observations of our world phenomena: (a) eighteen elements (Section 2.3) and (b) five aggregates (Section 2.4). The third observation of the world, i.e., *twelve links of interdependence*, is given in Chapter 4. It is rather astonishing to see that no serious contradiction is found between the ancient approach of *abhidharma* and the modern approach of brain science, in view of the fact that the former was a philosophy more than two thousand years ago. In Chapter 3, some findings of brain science are presented by comparing the two approaches of *abhidharma* and modern science and considering whether those are consistent or contradicting.

Detailed accounts on the five aggregates (*skandha*s) are given in Section 2.4. Observing the aggregates and all the phenomena that are arising within oneself and ceasing continuously without interruption, one realizes that there is nothing among them to be said definitively as an eternal self (cf. the timewise, empty nature in Section 2.6).

Vasubandhu was an earnest scholar during his youth in Kashmir who deeply studied the *Vaibhāṣika* teaching, which is a comprehensive and systematic philosophical edifice of the early Buddhist schools. He used to give lectures on Buddhism to the general public. During the day he lectured on *abhidharma*, and in the evening he distilled the day's lectures into a verse (*kārikā*). When collected together, those six hundred verses (*kārikā*s) gave a thorough summary of the entire system. He gave this work the title *Abhidharmakośa* (Treasury of *Abhidharma*). Later, Vasubandhu converted to Mahāyāna Buddhism under the influence of his half-brother Asanga, elder to him by twenty years. A source places Vasubandhu between c. 350 and 430 CE.

The following are some essential parts taken from his masterpiece *Abhidharmakośa-bhāṣya*, describing the world phenomena (dharmas), discerned on the basis of our six sense organs (six faculties).

2.3 Six Sense Organs (Six Faculties, 六感機能)

The *Abhidharmakośa* [2, 3] covers every subject of all phenomena (all dharmas) of our world and defines various philosophical concepts in order to capture them. The descriptions are basically scientific. Observing our experiences every moment, all phenomena are changing continuously, arising and ceasing, and being captured by the five aggregates, i.e. form, sensation, conception, mental formation, and consciousness (Section 2.5). The word *aggregate* means a collection of something; it means "made up of." It is used in a casual statement as: "The meeting was an aggregate of three Indian musicians and two Japanese fine artists." From a Buddhist point of view, aggregates are formed by causes or conditions coexisting in assemblage and interpreted as a composite. Composite phenomena are termed as aggregates.

The first aggregate, i.e. the form aggregate (*rūpa* aggregate), denotes materials in both external and internal worlds. Externally, *rūpa* denotes the natural world surrounding us, observed physically and captured in scientific ways. Internally, it denotes our body and physical sense organs. The sense organs are categorized into six faculties: eye, ear, nose, tongue, body, and mental faculty (i.e., brain). See Table 2.1.

The first five organs are called primary sensory organs. These five sensory organs are exposed to the external world, while the sixth, mental faculty, is housed internally in the skull. Because the function of brain or functions of internal organs in general were not known well two thousand years ago, the traditional *abhidharma* theory defines the mental faculty differently, as described in part (v) in Section 2.5. Three different names (mind, consciousness, and thought) designate the same referent object, i.e. the mental faculty (*manas*), but have different meanings: thinking, cognizing, and collecting together. They have a base: mental faculty.

It is proposed here that the sixth mental faculty is the brain, from a modern point of view. Any damage to the ear gives rise to trouble in hearing. In this way, one understands the ear as a hearing faculty. Similarly, any damage to the brain causes difficulty in thinking. In this way, one identifies the brain as a mental faculty. In fact, damage to a certain part of the cerebral cortex in the brain results in a language disorder known as Wernicke *aphasia*. An individual with Wernicke aphasia has difficulty understanding language. In modern brain science, there are a number of examples that indicate different mental disorders being connected to damage to certain different parts of the cerebral cortex. Thus, one understands the brain as a mental faculty.

The human cerebral cortex is shown in Fig. 2.1. The cerebral cortex receives input from the primary sensory organs and sends outputs to various parts of the body. Each of the signals captured through the primary sensory organs is projected (transmitted) to different interior sensory areas (cortices) in the brain (Fig. 2.1). The sensory areas are identified mostly in the rear part of the brain, and the remaining parts are identified as processing the perceived information and thoughts. The cerebral cortex

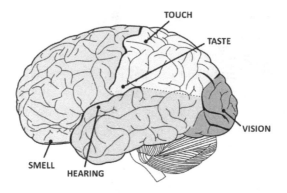

Fig. 2.1   Human *cerebral* hemisphere.

acquires information from the surrounding world, stores it, and processes it in various ways. Not only does the brain receive input signals, but it functions in a self-exciting way, day and night, twenty-four hours a day; the evidence of this is dreaming. Sometimes (always, according to some Buddhists), the brain misinterprets. This causes various defilements, investigated philosophically.

2.4 Eighteen Elements: Six Sense Faculties, Six Objects, and Six Consciousnesses

Before investigating the five aggregates described in detail in the next section, let us consider the fundamental elements of our beings, with which the treatise *Abhidharmakośa* starts. In its first chapter, eighteen elements (18 *dhātus*, 十八界) are defined, comprising six internal sense faculties (described in the earlier section), corresponding six external sense objects, and associated six consciousnesses. These eighteen elements function through the five aggregates. This is the ancient Buddhist view on our beings. The eighteen *dhātus* can be arranged into six triads, where each triad is composed of a sense faculty, a sense object, and sense consciousness (Table 2.1).

In the traditional Buddhist view, the number of sense organs is six, unlike the usual identification of five organs. In the Buddhist view, the sixth, i.e. mind, is regarded as an internal sense organ, which is

Table 2.1

	Faculty	**Object**		**Consciousness**
(a)	Eye	Visible forms	Eye-consciousness	(Seeing);
(b)	Ear	Sound signals	Ear-consciousness	(Hearing);
(c)	Nose	Odorous objects	Nose-consciousness	(Smelling);
(d)	Tongue	Taste objects	Tongue-consciousness	(Tasting);
(e)	Body	Touch objects	Body-consciousness	(Touching);
(f)	Brain	Signals from sensory faculties and thinking	Mind-consciousness	(Thinking).

reasonable. Among the mental objects, mental activities such as planning, motivation, and volition are included; additionally, various afflictions and desires are also included. This is central in Buddhism; hence, it is worth investigating from the viewpoint of modern brain science. Buddhist observations are very deep, although they were formulated about two thousand years ago. This will be seen in Chapter 3 through a comparison with modern brain science.

2.5 Five Aggregates (Five *Skandha*s, 五蘊)

Observing our daily life and ourselves every moment, we experience continuous change in both ourselves and the external phenomena around us. The external world and our perception are changing their appearance, arising and ceasing one after another, as if they are waves or bubbles appearing in rapid streams. Similarly, our minds too are continuously changing with the emergence and cessation of thoughts. In old times, all of these dharmas (i.e., all phenomena either internal or external) were the subjects of investigation of Buddhists as well as Indian scholars or *paṇḍitas*.

From a Buddhist point of view, these dharmas are interpreted either as composite (*saṃskṛta*, 有為) or as noncomposite (*asaṃskṛta*, 無為). The composite dharmas are called aggregates because they have been formed by causes or conditions coexisting in assemblage. The Sanskrit word *saṃskṛta* means "that which is formed or made by causes and conditions coming together and coexisting in assemblage." Because of this meaning, the word *composite* is used here for the Sanskrit word *saṃskṛta*, although some literature (e.g. [2]) uses "conditioned" instead of composite.

Full discernment of dharmas is essential to attain liberation from suffering. This is stated at the beginning of the treatise *Abhidharmakośa* [2, 3]. It states:

> *Without full discernment of dharmas, there is no method to pacify or appease the afflictions. It is the afflictions which cause the people to wander in the ocean of existence.*
>
> (*Kārikā* No. 3, Chap. 1 [3]).

Kārikā 7 (of Chap. 1 [3]) interprets composite dharmas as:

> *Composite dharmas are the five aggregates (skandhas).*

The *skandhas* (Sanskrit) or *khandhas* (Pāḷi) are the aggregates (composite collections) that constitute living beings and ourselves. The five aggregates are:

Forms (*Rūpa-skandha*, 色・形蘊),
Sensation (*Vedanā-skandha*, 受蘊),
Conceptions (*Saṃjñā-skandha*, 想蘊),
Formations (*Saṃskāra-skandha*, 行蘊), and
Consciousnesses (*Vijñāna-skandha*, 識蘊).

These are defined as the five factors that constitute and explain a sentient being and personality. Aggregate means a collection of something; it means "made up of."

The five aggregates are (i) *rūpa*: form (our body, sense organs, and/or external objects); (ii) *vedanā*: sensations (received by sense organs); (iii) *saṃjñā*: conceptions (resulting from the sensation process); (iv) *saṃskāra*: formations (mental activity associated with the conception); and (v) *vijñāna*: consciousness (overall cognition as apperception).

It is central for Buddhists to realize that suffering arises when one clings to the aggregates. By relinquishing the clinging or attachment to the aggregates, suffering subsides and is extinguished. Although the five aggregates contribute to the rise of craving and clinging, Buddhists have the profound view that the nature of all aggregates is intrinsically empty as an independent existence (Chapter 9: "Nāgārjuna's Philosophy on *Śūnyatā*"). According to the deep insight on the intrinsic emptiness-existence, the appeasement of craving and clinging to the aggregates is realized by following practices of the Buddhist tradition.

Each of the five aggregates are explained below in detail according to the *Abhidharmakośa*. Surprisingly, these are consistent with the findings of modern brain science, which are given in the next chapter.

(i) *Aggregate of form* (*rūpa-skandha*, 色蘊)

The aggregate of form (*rūpa*) denotes the forms of both internal and external materials. Internally, *rūpa* denotes our body and physical sense organs, and externally it denotes the natural world around us. In particular, the five sense organs, i.e., the eye, ear, nose, tongue, and body, are associated with the corresponding five external *rūpas* (referents) of visible forms, sounds, scents, tastes, and tangibles. The tangibles include the natural environmental objects such as air, water, and earth, in addition to the direct sensations of touch (such as soft, rough, cold), hunger, and thirst.

Needless to say, these are all subjects of scientific study. Hence, modern scientific studies provide ample evidence for these interactions.

Linguistically, the word *rūpa* has a twofold meaning: (i) that which is subject to deformation, breaking, or resistance, and (ii) shape (形) or color (色).

(ii) *Aggregate of sensation* (*vedanā-skandha*, 受蘊)

The aggregate of sensation is associated with sensory experience of external objects and is divided into six groups resulting from contact of the six sense organs (eye, ear, nose, tongue, body, brain) with external signals or objects (light, sound, odor, taste, touch, minding). The Sanskrit word *vedanā* is often translated into English as feeling or sensation, i.e., the ability to sense something physically, or a physical sensation that results from associated faculties.

Chapter 3 describes physical sensations studied by brain science. Here, the term *sensation* is chosen, owing to its direct meaning rather than the feeling, because it seems to capture the momentary nature of *vedanā* better and because feelings in ordinary parlance get mixed with emotions and mental events, which fall more into the fourth aggregate, formation [2].

(iii) *Aggregate of conception* (*saṃjñā-skandha*, 想蘊)

The function of conception is to discern individual fine distinctions and to perceive, such as "this is blue," or "yellow, long, short, agreeable, disagreeable," etc. The aggregate of conception is the seizing or apprehending of attributes of the sensation or sensed objects. A sensation is like a physical function of a tool of scientific measurement, while conception is a kind of human function to apprehend the meaning of sensed objects, such as the attributes of sensed objects. Hence, it is an initial stage of mind formation.

Consider visual consciousness. When a signal of the color blue (say) is sensed as an impression gained from an external object, the subsequent step of perceiving is taken, such as "The object is blue." Therefore, it is through conception that a name is given to a visual impression.

With regard to perceptions, such as blue, yellow, long, short, etc., the aggregate of conception (*saṃjñā-skandha*) comprehends them by conceptually combining together their appearances, names, and signs (male, female, etc.). It becomes the cause of reasoning and investigation. Thus, it is named conception (*saṃjñā*). It is not a simple perception, but concept formation. Names or words, such as color or sound, are understood to be those that constitute the conception. Name and concept are identified with one another.

Saṃjñā is the grasping of an object's appearance. The object is the cognitive object. Its appearance is what causes the cognitive object to be assigned a blue, yellow, or other color. The grasping of this appearance is the determination that "this is blue, not yellow."

The Sanskrit *saṃjñā* is usually translated as "perception." Its Sanskrit form, *saṃ-jñā*, is composed of the prefix *saṃ-*, "together," and the root verb *jñā*, "to know, perceive, understand," that is, a "together-knowing." *Saṃjñā* thus means "conception, idea, perception." *Saṃjñā* (the third *skandha*) is etymologically the opposite of *vijñāna* (the fifth *skandha*), which is composed of *vi-*, "dis-," and the same root *jñā*, "to know." While *vijñāna* (consciousness, or recognizing) stresses disjunctive (hub-like, see(v)) discernment, *saṃjñā* emphasizes a conjunctive construction of an image or idea that brings disparate sensations together into one, often connected with a name or concept. *Saṃjñā* is an initial stage of mind formation, which develops into mental formation in the subsequent stage of the aggregate of formation.

In real life, we often experience various perceptions, such as pleasant, unpleasant, satisfactory, unsatisfactory, and equanimous (see Sections 3.3(a) and 3.6(a)). A pleasant experience is that which one desires to be united with again when it ceases. An unpleasant experience is that which one desires to be separated from when it arises. An equanimous experience is that which is neither a thought nor an emotion, but rather a steady, conscious realization of reality's transience. These human emotional experiences are controlled by the physiological function of neurochemicals, such as dopamine, noradrenaline, and serotonin, presented in Section 3.3(a).

(iv) *Aggregate of formation (saṃskāra-skandha, 行蘊)*

The aggregate of formation means the conditioning forces, which are characterized by mental formation, intention, or volition and which provide motivation for an action (karma). Volition means the consciousness of choosing or determining. This is of capital importance since it creates and determines future developments. The aggregate of formation is interpreted as constructing activities or karmic activities, including all types of mental imprinting and conditioning triggered by an object. It includes any process that makes a person initiate action or acts.

Emotions may have an impact on action generation. This is a notable aspect of emotions, influencing decision making or choosing between alternatives. Decision making is the process of choosing a preferred option or course of action from a set of alternatives (see Section 3.6 (b)). If the initial conception was mistaken in capturing reality, the resulting mental formation would be stained.

There are several classes of volition (intending to do): volition regarding forms, hearing, tactical actions, and mental formation. The aggregate of formation is composites (*saṃskṛta* 有為) that differ from the other four aggregates: form, sensation, conception, and consciousness.

(v) *Aggregate of consciousness* (*vijñāna-skandha*, 識蘊)

The aggregate of consciousness is that which makes known, i.e., apperception, namely the cognition of the object field as a whole (*Kārikā* 16 of Chap. 1). Apperception means the process by which perceived qualities of an object are related to past experience. We get a perception of a chair through our eyes, but apperception is how our mind relates it to chairs we have seen before.

Consciousness is not simple. There are six groups of consciousness: visual, auditory, gustatory, olfactory, tactile, and mental. For example, visual consciousness apprehends color (blue, etc.) or shape; it is associated with a certain perception, which apprehends a certain characteristic: "This is a man, this is a woman." Mental consciousness apprehends: "This is the blue coat of the girlfriend." This is a kind of *self-consciousness* and the apperception of the fifth aggregate of consciousness (*vijñāna-skandha*).

Consciousness is believed to be associated with the hub-like function of the organ named thalamus (chamber), which belongs to the limbic system in the brain (as seen in Section 4.3 (c)). The thalamus plays a major role in regulating arousal (level of awareness) and activity. Thalamic nuclei have strong, mutually reciprocal connections with the cerebral cortex, forming thalamo-cortico-thalamic circuits that are believed to be involved with consciousness. It is amazing to realize that the ancient analyses of *abhidharma* theory cited so far resemble the understanding of the brain structure revealed by modern brain science.

There is no mental faculty that is visible externally. According to modern brain science, the brain undoubtedly participates in the aggregate of consciousness. One of the locations is identified at the Thalamus (Section 3.4 (b)). The mental faculty is nothing but the brain within the skull, but its function is not restricted to consciousness only (see (vi)).

The processes of consciousness succeed one after another sequentially, associated with five sense organ spheres (visual, auditory, olfactory, gustatory, and tactile) and with the sixth, mental sphere. The consciousness that has just disappeared is regarded as a preceding condition for a subsequent one and becomes the basis of the consciousness that immediately follows. This is the reasoning and interpretation for consciousnesses according to the *abhidharma* theory [2]. Because of this aspect, it (the preceding consciousness) receives the name of mental faculty. Furthermore, the theory states that the aggregate of consciousness is the sense sphere of the mental faculty, and that its function has seven elements: i.e., in addition to the six elements of consciousness noted above, there is the element of mental faculty (*manas*). Thus, the *abhidharma* philosophy proposes the existence of the mental faculty, *manas*. Clearly, the first five consciousnesses have the five sense faculties, i.e., eye, ear, nose, tongue, and body for their bases, with the sixth being brain (*consciousness citta*) and the seventh, Manas (mental faculty).

Mind (*manas*), thought (*citta*), and consciousness (*vijñāna*) are the three different names that designate the same referent object, i.e. the mental faculty (*manas*), but have different meanings: considering or thinking, collecting together, and cognizing, respectively. They have a base: the mental faculty. *Manas*, *citta*, and *vijñāna* are each sometimes used in the nontechnical sense of "mind" in general, and the three are sometimes used in sequence to refer to one's mental processes as a whole.

From a modern point of view, it is very hard for us to understand what is meant by *manas*. However, more than two thousand years ago, it should have been very hard for the ancient people to capture the function of the brain without knowing how the neural networks that run throughout the body are connected with the brain and how they work together.

(vi) *Manas* and *ālaya vijñāna*
At the same time when Vasubandhu was writing the *Abhidharmakośa,* Mahāyāna Buddhism was developing at the same place, Gandhāra. As stated in Section 2.1, Vasubandhu later converted to the Mahāyāna under the influence of his brother Asaṅga (c. 330 to 405 CE), who was one of the important figures of Mahāyāna Buddhism, as well as Maitreya-nātha and Ācārya Nāgārjuna. The Yogācāra (literally "yoga practice," "one whose practice is yoga") is an influential school of Buddhist philosophy emphasizing the study of cognition, perception, and consciousness through meditative and yogic practices. It is said that the systematic exposition of Yogācāra philosophy owes much to Asaṅga and Vasubandhu. However, actually, the philosophical school (Nagarjuna school) had already existed during their times.

The most famous innovation of the Yogācāra school was the doctrine of eight consciousnesses. Yogācāra philosophy reformulated the *abhidharma* philosophy of consciousnesses into its own schema, with three novel forms of consciousness in addition to the five consciousnesses of *Abhidharmakośa*. The sixth consciousness, *mano-vijñāna*, was seen as the surveyor of the content of the five senses as well as of mental content such as thoughts and ideas. The seventh consciousness, *manas-vijñāna*, developed from the early Buddhist concept of *manas* and was seen as the defiled *manas* (*kliṣṭa-manas*), which is obsessed with notions of self. The first five consciousnesses are associated with five sense organs, the sixth is the *citta* (mentality), the seventh is the *manas* (self-consciousness), and the eighth is the storehouse or substratum consciousness.

The thinking done by *manas* is more closely linked to volition than to the discursive processes associated with the apperception thinking of *citta*. The *manas* is mainly the mental activity that follows from volitions, caused by the activation of a latent tendency. The first six consciousnesses are the ones without substantial entities (empty of entities) but with a series of events, arising and vanishing, although there are associated sense organs. Consciousness is different, and it is in fact a spatiotemporal event. This new formulation may be connected with the Vasubandhu's conversion to the Yogācāra Mahāyāna school.

The seventh consciousness, *manas-vijñāna*, is captured as the defiled *manas* that is obsessed with notions of self, while the eighth is called *ālaya-vijñāna*, or *ālaya*, storehouse. Both truth and defilement, principle and phenomena, or ultimate way and nescience (see Section 9.3.2) are interpenetrated in *ālaya-vijñāna*. This is the view of the awakening in the Mahāyāna [12]. Philosophical unification of "not identical condition" on the one hand and "not different condition" on the other hand was the main theme of Nāgārjuna's *Treatise of the Middle* (*Madhyamaka-kārikā*), which is given in Chapter 7. This Mahāyāna philosophy was already proposed in the first half of the 2nd century CE.

Note: Yogācāra philosophy holds that being unaware of the processes going on in *ālaya-vijñāna* is an important element of the ignorance (nescience, *avidyā*) of human beings. At the root of human existence is the nescience, that is the fundamental view of Buddhists.

(vii) *How we are described by the aggregates?*
Let us consider how our daily experiences are described by the five aggregates with the help of two examples. The first is a case of cooking food:

(i) Consider a kitchen and our sense organs. These are the first aggregate: form; (ii) The sensation of smell is caught by the nose faculty. This is the second aggregate: sensation; (iii) Grasping of the smell follows with such thoughts as "This is a curry smell, not other." This is the third aggregate: conception. Thus, the odorous impression is identified and named. Suppose that curry spice is one of the favorite foods that one desires to have. (iv) Fourth is the aggregate of formation, stimulating the motivation for action. In other words, the smell of curry spice stimulates the passion for cooking one's favorite food; (v) One begins to plan how to cook this food using curry seasoning and is reminded of past experiences. This is the fifth aggregate: consciousness. Now it is the apperception of how the mind is related to the cooking that one has experienced before.

Consider another example: In the course of cyclic existence (*saṃsāra*), men and women are mutually enamored of each other's forms (*rūpa*) (i), which they find attractive and charming. This is the sensation aggregate (*vedanā*) (ii). Trying to apprehend what they are feeling is the conception aggregate (*saṃjñā*) (iii). Their conceptions become the cause of subsequent reasoning and investigation, which is the fourth aggregate of mental formation (iv).

Conception is an initial stage of mind formation, and subsequent mental formations are diverse — wise, unwise, or equanimous — in a steady sequence of thinking. However, sometimes, misconception happens, provoking careless attachment. This causes one to fall into unwanted troubles, sometimes emotional. Attachment is a driving force of the formation of karmic action (*saṃskāra*). (v) Consciousness (*vijñāna*) is the thought of what they were trapped by or what they have done. These thoughts are inevitably often illusionary. From the Buddhist point of view, those are *kleśa* or defilements. In comparison, the former example is equanimity, not like an emotion, but rather like a steady conscious realization of reality's transience.

Why is there misperception, mistake, illusion, or defilement?

We briefly consider what each of the aggregates of sensation, conception, and formation means in Buddhism.

Sensation and conception are two aggregates in the following ways: (1) Sensation and conception are two roots of attachment to pleasures and attachment to views; (2) Those are two elements of the *saṃsāra* existence of cyclic linkage. The second issue will be considered in Chapter 4. Here, the first issue is taken up. In fact, if one becomes attached to pleasures, it is because one enjoys sensation; if one becomes attached to a view, it is a false view or a mistaken conception, actually a misperception.

According to the *abhidharma* theory [2], sensation and conception are causes, giving rise to two types of defilements: desires and (false) views. Desires arise through the power of sensations, while views arise through the power of conception. All defilements have these two as their spring-heads: sources of false views.

There are four mistaken views according to the *abhidharma* theory. Such false views include all the defilements to be abandoned by the path. The four mistaken views are:

 i. Taking that which is impermanent to be permanent;
 ii. Taking that which is unsatisfactory to be satisfactory;
iii. Taking that which is impure to be pure;
 iv. Taking that which is not a self to be a self.

False views are to be abandoned, because from them proceed wrong thought, wrong speech, wrong action, and wrong livelihood.

Ignorance, craving, and grasping are defilements in their intrinsic nature. *Kārikā* 36 of chapter 3 [2] states: "Defilement is explained to be like a seed, a root, and a *nāga*." From the seed, there arise a sprout, stalk, leaves, etc. In the same way, from defilement there arise misperception, action, and phenomenal basis. The pond where *nāgas* (snakes, cobras, etc.) live does not dry up. In the same way, the ocean of births where defilement lives does not dry up.

It should be tremendously interesting to consider any relevance of the five aggregates to modern brain science. In fact, the aggregates of sensation and conception can be related to two different parts of the brain (mental faculty). Similarly, formation (fourth aggregate) and consciousness (fifth aggregate) are interpreted by the functions of the brain. These are considered in Section 3.6, in the next chapter.

2.6   Five *Skandhas*: Receptive to Emancipation, Resulting in Liberation

Phenomenologically speaking, we can think of human beings as living in sequentially changing states of *skandhas*, which arise continuously and cease from moment to moment of our existence. In the *abhidharma* theory, it is essential to understand that the skandhas have an evolving (transitional) nature of phenomena (dharmas) or beings. This is analogous to the mechanical laws of modern science, which describe dynamical evolution of mechanical systems.

The five skandhas are nothing but the composites (i.e., those dharmas formed or made through causes and conditions). According to the scholars of *abhidharma* [2, 3], the composites (*skandhas*) are said to be timewise or paths. The composites are called timewise because they arise in three times: past, present, and future, i.e., they have arisen (or ceased), are arising (or ceasing), or will arise (or cease) in the three times, respectively. Or, they are called timewise, because they are changeable. Additionally, the composites are like paths by which people went to, are going to, or will go to the market in the three times. These views remind us of the orbits of dynamical evolution of a mechanical system in physics. According to a mechanical law, a body with a mass moves along an orbit (a path) with respect to time, ranging from past through present to future. In this way, the composites are timewise and changeable, not static.

Owing to their timewise or changeable nature, the composites are regarded as unsubstantial, void, or empty in nature. From the point of view of Mahāyāna Buddhism, the composites are receptive to emancipation (emancipating oneself physiologically from suffering). Once one transcends these composites, *nirvāṇa* will be attained, resulting in liberation. Without full discernment of the dharmas, there is no method to attain liberation. Because of afflictions, the worldly beings accumulate karma and they do not achieve freedom.

It is essential to realize that the *saṃskṛta* dharmas (composites, five *skandhas*) exist in the form of compounds that are transforming themselves ceaselessly. This is the only way they exist. Dharmas of this world are dynamical events rather than static events and are character-ized as essentially changeable in nature. Our mind is also composite and essentially changeable in nature. This very fact enables emancipation (liberation) of the mind from afflictions. This is the essential view of Mahāyāna.

Chapter 3

Modern Brain Science and Ancient Buddhist Concepts on Mind

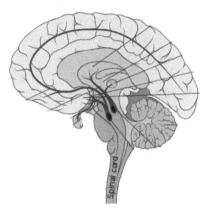

Introduction

Modern sciences have been disclosing a number of new findings on our brain functions. The brain receives information from the external world, interprets it, and embodies it into actions in the real world. In addition, it controls the functions of various internal organs of the body. The brain receives external information through sense organs such as the eye, ear, and others. It assembles information in a way that has meaning for our body, stores it as our memory, and controls our thoughts, memory, movement of arms and legs, and so on. The brain is composed of the cerebrum, cerebellum, and brain stem, protected within the skull. The surface of the cerebrum is called the cortex, which has a folded appearance with hills and valleys.

It is remarkable to see that most observations of the ancient *abhidharma* philosophy presented in Chapter 2 have many common features observed by modern brain science. For example, five types of perception (*vision, sound, smell, taste*, and *touch*) are captured through the five sense organs (five faculties): eye, ear, nose, tongue, and body. In addition, the traditional *abhidharma* theory defines a mental faculty that would be attributed to certain functions of the brain by the modern view.

In addition, the *abhidharma* philosophy states that our existence is basically characterized by five aggregates. In other words, living beings are characterized by five functional assemblages: *forms* (denoting internal and external materials), *perception, conception, formation*, and *consciousness*. Modern science categorizes perceptions as exteroception (external sensory system) and interoception (internal sensory system). In fact, the function of the latter is implicitly the ultimate target of Buddhists. Actually, Buddhists take practical approaches to it. The second half of this book is devoted to elucidating the profound approach and philosophy of Buddhists associated with interoception.

Modern science attributes the aggregates to the structure and function associated with the human brain, spinal cord, and neural networks that run throughout our body. The findings captured by Buddhist practices historically over long terms are consistent with the findings of the modern sciences, such as neurobiology and brain science. Remarkably enough, according to a recent study, certain neurobiological effects of the Buddhist practices of meditation and mindfulness have been investigated by scientific methods successfully.

Needless to say, modern brain science has disclosed a number of mechanisms, which ancient people would never have imagined. Some of them are the function and structure of the brain itself, neural networks, neurotransmitters, etc. They are presented here in Sections 3.1–3.4 and Appendix A and compared with the *abhidharma* theory.

3.1 Sensory Areas in the Brain

Our brain receives a bunch of signals with our sensory organs. Each of the signals captured through the five sensory organs (eye, ear, nose, tongue, and body) is projected to each separate sensory area (cortex area) in the brain. Human cerebral cortex is shown in Fig. 3.1. These sensory areas are identified mostly in the rear part of the brain, while the cortices associated

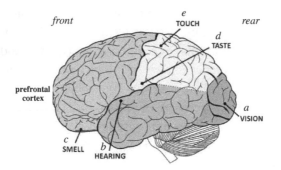

Fig. 3.1   Human cerebral hemisphere.

with body movement and voluntary planning are located in the front part. In the figure, sensory regions that receive signals from external objects are shown for each receptor cortices of sight, sound, smell, taste, and touch: (a) vision cortex, (b) hearing cortex, (c) smell cortex, (d) taste cortex, and (e) touch cortex.

Visual signals are first perceived by the *retina* in the eye bulb, as shown in Fig. 3.2, and then the signals enter the visual cortex *a*, which is interconnected to neighboring areas for further processing. The auditory feeling is the perception by the *cochlea*, and then the signals enter the auditory cortex *b*, which is interconnected to neighboring areas. Taste and smell are perceived by the taste buds on the tongue and smell receptors (olfactory bulb) on the bottom side of the brain. These signals enter the gustatory cortex *d* and olfactory cortex *c*, respectively, for further processing (see Fig. 3.3). The somatosensory (touch or tactile) feeling is perceived by a system of nerve cells that responds to changes to the surface or internal state of the body. The signals are sent along a chain of nerve cells to the *spinal cord*, where they may be processed by other nerve cells and then relayed to the somatic (body) sensory cortex *e* in the brain for further processing.

Thus, the sensory parts have been summarized above. The frontal lobe (Fig. 3.8) plays a large role in voluntary movement. The primary motor cortex (Fig. 3.5) regulates motion, such as walking. Walking is not a simple mechanical motion, but an organized motion controlled by the brain.

The function of the frontal lobe involves the following: the ability to project future plans resulting from current state and action, the choice between good and bad actions (or better and best), and the determination of similarities and differences between things or events. The frontal lobe also plays an important role in retaining longer-term memories. These are memories associated with learning or emotions. The frontal lobe modifies those emotions to fit generally appropriate forms or socially acceptable norms. Sometimes, this modification causes, often accidentally, erroneous images or illusions.

The frontal lobe (Fig. 3.8) contains most of the *dopamine*-sensitive neurons in the cerebral cortex. The dopamine system is associated with *attention*, *planning*, *motivation*, *volition*, and *long-term memory*, which are regarded as matters of the mind anyway.

3.2 Visual Perception

Seeing an object is the perception represented by a chain sequence of signal transductions shown in Fig. 3.2. The visual system is regarded as a *stimulus-response* system. In this figure, the stimulating light ray coming from the object ① excites electrical response in the visual cortex portion ④ of the brain. The network of the system is compound and contains many subsystems, each of which can be viewed as a stimulus-response system.

The system connecting retina ②, LGN ③ (small dark oval, lateral geniculate nucleus), and visual cortex ④ can each be viewed as a

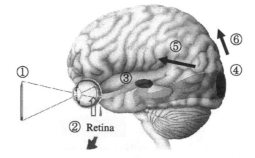

Fig. 3.2   Visual system: ① Visual object; ② eye retina; ③ LNG; ④ primary visual cortex; ⑤ ventral stream; ⑥ dorsal stream.

stimulus-response system. The signals filtered at the LGN ③ enter the primary visual cortex ④, which performs certain primary processing of the filtered nerve signals. After the processing, the nerve signals are transmitted to the secondary visual cortices, where these signals are interpreted.

Here, the brain participates in signal processing someway. During initial reception, raw images are sampled by the retina ②, each of which provides a very limited scope of vision of the object seen. In addition, the images on the retina are only raw spots or dots that photons (discrete signals) of the light ray generate there. Using the raw images, a whole mental image is created in the brain by integrating the retinal images perceived by multiple fixations of different unconscious focusings. Thus, the visual image we perceive is a *mental construction* inside our brain. The mental construction is so real and compelling that we rarely question its reality.

Each image of the primary visual cortex transmits information to two primary pathways, called the ventral stream ⑤ and the dorsal stream ⑥. Thus, visual labor is divided between *vision-for-perception* and *vision-for-action*, namely *what* pathway and *where-how* pathway. The former is associated with form recognition and object representation, while the latter is associated with object motion, object locations, and self-control of the eyes and arms.

At each moment, we perceive a very small fragment of the world through the retina, yet our subjective perception of the visual world in front of us is rather clear, coherent, and complete. Often, we see things that are not even there. This is because what we perceive is actually a *virtual visual world that is created in our mind*, which is a *mind-product* of interaction between the *incoming sensory data*, our *experience*, and *prior knowledge*. The last is what is meant by the fifth agrregate of consciousness (Section 2.5 (v)).

3.3 Mental Faculty and Internal Sense Faculties

According to the *abhidharma* theory, the aggregates (conception, mental formation, and consciousness) are associated with the mental faculty (Section 2.5 (vi)). According to modern science, it might be said that the mental faculty is located within the brain.

Human sensing is a process of gathering information about the surrounding world to respond to a stimulus. Regarding the human body, the brain receives signals from the sense faculties (organs), which continuously receive information from the environment, interpret those signals, and cause the body to respond, either physiologically or physically. Five human perceptions are described as *seeing, hearing, smell, taste,* and *touch.* During sensing, sense organs collect various stimuli (such as voice or smell) for signal transduction, i.e. transformation into a form that can be understood by the brain. Sensing and conception are fundamental to every aspect of human cognition, behavior, and thought. *Cognition* refers to mental action, acquiring knowledge and understanding through thought, experience, and the senses.

A sense faculty such as an eye consists of a group of interrelated sensory cells that respond to a specific type (a modality) of physical stimulus such as color signals. Sensory modalities represent how sensory information is encoded or transduced in different ways. The sensory modality of vision is *light* or *color.* To perceive a light stimulus, the eye must first refract light so that it directly hits the retina (Fig. 3.2). The transduction of light into neural activity occurs via the photoreceptors in the retina. Multimodality, such as *light, voice,* and *smell,* integrates different senses into one unified perceptual experience such as a scene. For example, information from one sense has the potential to influence how information from another is perceived.

Through brain neurons and spinal nerves (nerves of the central and peripheral nervous systems), sensory information is relayed to and from the brain and body. Different types of sensory receptor cells in sensory organs transduce the sensory signals from the organs toward the central nervous system, finally arriving at the sensory cortices in the brain, where sensory signals are processed and interpreted (perceived) such as a scene. Such a group sensing is called an aggregate by the ancient *abhidharma* philosophy.

Sensory systems, or senses, are often divided into external (exteroception) and internal (interoception) sensory systems. Human external senses are based on the sensory organs of the eyes, ears, nose, mouth, and skin. Internal sensing detects signals from internal organs and tissues. Internal senses possessed by humans include the *vestibular* system in the inner ear (sensing body balance and spatial orientation), as well as *proprioception* (sense of body movement and posture), and *nociception* (pain, detection of potentially damaging mechanical or thermal stimuli). Other internal senses are the signals, such as *hunger, thirst, suffocation*, and *nausea*, or different involuntary behaviors such as *vomiting*.

In Buddhist philosophy, in addition to the five external sense organs, the *āyatana* (sense base) includes the mind as a sense organ. This addition to the five sensory organs represents a remarkable approach of Buddhists. Apparently, the mind is considered as a different spectrum of phenomena that differ from the physical sense data. However, this way of viewing the human sense system indicates the uniqueness of the Buddhist philosophy, recognizing the importance of internal sources of sensation and perception (i.e. interoception) that complements our experience of the external world. Modern science is making it clear that this is nothing but the Buddhist approach to the whole world, both internal and external, which will be seen in the following sections.

3.4 Brain Areas: *Synthesized Views, Consciousness,* and *Emotion*

(a) Brain area related to *synthesized views: Cerebrum*

According to the *abhidharma* philosophy, the aggregates of conception and mental formation grasp the perceived objects by conceptually combining together their appearances, names, and the signs (male, female, etc.). They comprehend the appearance of an object and perceive that this is agreeable or disagreeable. This is the cause of reasoning and investigation. Names or words, such as *red color* or *voice*, are understood to cause conception to arise. The aggregate of formation is interpreted as a *constructing activity*, including all types of mental activity triggered by an object and making a person initiate action.

According to modern brain science, the cerebrum (Fig. 3.1) functions as the center of sensory perception, memory, thoughts, and judgment; it also functions as the center of voluntary motor activity. As mentioned in Section 3.1, the primary sensory areas receive and process visual, auditory, olfactory, gustatory, and somatosensory information.

Combining those with associated cortical areas, the cerebrum synthesizes sensory information into our view, i.e. synthesized view of the world around us.

Very often, during this phase, perceptions are not always correct and some mistaken perceptions occur. As considered in Section 2.4 (vii), one becomes attached to mental formations, and mistaken views arise from them, together with the emotional sensations.

Recent brain studies have revealed that "holding something in mind" is associated with activity in an extended brain area involving both prefrontal cortex and more posterior areas. In the prefrontal cortex (Fig. 3.1), highly processed information projected from various sensory cortices is integrated in a precise fashion to form a memory, perception, and intricate action. Diverse cognitive processes are monitored here.

Thus, general synthesized views around oneself are formed in the prefrontal cortex.

(b) Brain area related to *consciousness: Thalamus*

In the brain, there is a hub-like place for various signals from the sensory systems. The place is named thalamus (chamber) (②) of Fig. 3.3) and has

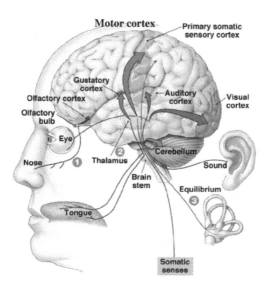

Fig. 3.3   Cerebrum, Thalamus, Cerebellum and Brainstem

multiple functions. It is generally believed to act as a relay between different subcortical areas and the cerebral cortex. This is associated with *consciousness*.

For hearing, for example, inputs from the cochlea in the inner ear are sent to the thalamic auditory nucleus (②) of Fig.3.3), which in turn projects to the auditory cortex. The thalamus is believed to both process sensory information and relay it. In this way, every sensory system includes a thalamic nucleus that receives the sensory signals and sends them to the associated primary cortical area. For the visual system too, inputs from the retina are sent to the LTN (lateral tegmental nucleus) of the thalamus, which in turn projects to the visual cortex. In addition, each of the relay areas in thalamus receives strong feedback connections from the cerebral cortex.

The thalamus plays a major role in regulating arousal (the level of awareness) and activity. Thalamic nuclei have strong reciprocal connections with the cerebral cortex, forming thalamo-cortico-thalamic circuits that are believed to be involved with consciousness.

The thalamus has been thought of as a "relay" that simply forwards signals to the cerebral cortex. However, thalamic function is more selective. Many different functions are linked to various regions of the thalamus. This is the case for many of the sensory systems (except for the olfactory system), such as the visual system, auditory, gustatory, somatic (body), and visceral (internal-organ) systems.

A major role of the thalamus is to support the motor system and the language system. The thalamus is functionally connected to the hippocampus with respect to spatial memory. Damage to the thalamus can lead to permanent coma. The thalamus also plays an important role in regulating states of sleep and wakefulness.

Thus, it is seen that the agrregate of consciousness (Section 2.5 (v)) is closely associated with the function of the thalamus.

(c) **Brain area related to** *emotion: Limbic system*

Our feelings are often characterized as pleasant, unpleasant, satisfactory, unsatisfactory, or equanimous (Section 2.5 (iii)). The first four may be termed as emotions in general. Here, emotions are reviewed biologically and neuroscientifically. It must be emphasized that emotions involve the entire nervous system. There are two parts of the nervous system that are especially significant: the limbic system in the brain and the autonomic nervous system. Human emotional experiences are controlled by the physiology of neurochemicals, such as dopamine, noradrenaline, and serotonin, which are important neurochemicals of the limbic system.

The limbic system is known as the area of the brain involved with emotion and memory. Its structures include the *thalamus, hypothalamus, amygdala,* and *hippocampus.*[1]

The following structures are considered to be related to emotion.

➢ *Thalamus* — The thalamus (chamber) is involved in relaying sensory and motor signals to the cerebral cortex, especially visual stimuli. The thalamus also plays an important role in regulating states of sleep and wakefulness. The thalamus is a hub-like place for various signals from the sensory systems (with the exception of the olfactory system). This place is named thalamus (chamber), and has multiple functions. It is generally believed to act as a relay between different subcortical areas and the cerebral cortex.

➢ *Hypothalamus* — The hypothalamus (under-thalamus) is located below the thalamus.

It plays a role in emotional responses by synthesizing and releasing *neurotransmitters*, which can affect mood, reward, and arousal. The hypothalamus is responsible for regulating hunger, thirst, response to pain, levels of pleasure, satisfaction, anger, and aggressive behavior.

It also regulates the functioning of the *autonomic nervous system*, which means that it regulates things such as pulse, blood pressure, breathing, and arousal in response to emotional circumstances (C. George Boeree, 2009).

➢ *Amygdala* — The amygdalae are two small almond-shaped groups of nuclei located deep within the temporal lobe of the brain, performing a primary role in the processing of emotional reactions, decision making, and memory. The amygdalae are involved in detecting and learning which parts of our surroundings are important and have emotional

---

[1] Source: "Biology of Emotion," Boundless Psychology , Apr. 13, 2016.

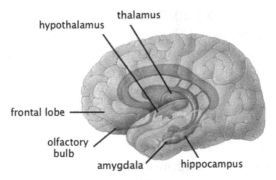

thalamus

hypothalamus

frontal lobe

olfactory
bulb

amygdala   hippocampus

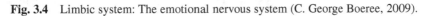

Fig. 3.4   Limbic system: The emotional nervous system (C. George Boeree, 2009).

significance. They are critical for the production of emotion, particularly, negative emotions such as fear. The amygdalae perform primary roles in the formation and storage of memories associated with emotional events. The amygdalae are also involved in appetitive (positive) conditioning.

➢ *Hippocampus* — The hippocampus is named after its resemblance to the seahorse, from the Greek. In the limbic system, it plays important roles in the consolidation of information from short-term memory to long-term memory and spatial navigation. Two hippocampi are located under the cerebral cortex, one in each side of the brain.

The hippocampus acts as a *cognitive map* — a neural representation of the layout of the environment. It is a frequent observation that without a fully functional hippocampus, humans may not remember where they have been and how to get where they are going: getting lost is one of the most common symptoms of amnesia (memory loss).

The limbic system, including the thalamus, hypothalamus, amygdala, and hippocampus, categorizes human emotional experiences as either pleasant or unpleasant mental states based on the physiological function of neurochemicals such as *dopamine, noradrenaline,* and *serotonin* (see Section 3.5). Other than emotion and memory, it supports the functions of motivation and learning as well. It is thought that emotions such as happiness, sadness, fear, anger, and disgust (and many others) are mental states that are constructed with many different systems in the brain working together (see Section 2.5 (iii)). There is evidence that the hippocampus also plays a role in finding shortcuts and new routes between familiar places. For example, London's taxi drivers must learn a large number of places and the most direct routes between them. A study showed that the part of the hippocampus is larger in taxi drivers than in the general public, and that more experienced drivers have bigger hippocampi.

(d) Action, movement, and emotion

Actions are usually embodied in voluntary bodily movement under the control of the brain motor system. The cerebral cortex plays a key role in memory, attention, perception, awareness, thought, language, and consciousness. Based on the present perceptual information, past experience, and future goals, nervous processing in motor cortex in the cerebrum is critical for action planning. The motor cortex (Fig. 3.5) and brainstem structures, with the assistance of the cerebellum (Fig. 3.3) and basal ganglia, translate an action intention into movement and coordinate the execution of an action plan. The brainstem contains most of the neural structures essential for rhythmic activities involving breathing, eating, eye movements, and facial expressions. In addition, the brainstem also projects to the spinal cord.

The *anatomical* hierarchy of the motor system supports the *functional* hierarchy of action organization. Abstract action intentions or plans

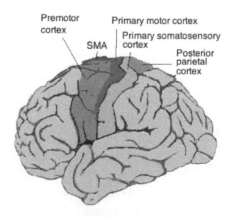

Fig. 3.5   Motor cortex

are formed on the highest levels in the prefrontal cortex (Fig. 3.1), transformed into motor programs at intermediate levels in the cerebral and cerebellum cortices (Fig. 3.3), and then implemented at the lowest bodily levels of the hierarchy.

Endogenous voluntary movement originates in the cerebral cortex (Fig. 3.1). The intentions, plans, or goals of action need not be concerned with the details of a bodily movement. Lower levels of mechanisms are needed to translate and realize motor commands into muscular movements. The primary motor cortex (Fig. 3.5) controls relatively simple and routine features of movement. The cerebellum and the basal ganglia provide feedback circuits that regulate cortical motor areas and the brainstem. They are necessary for smooth and accurate movement and posture.

Emotional perception (Section3.4 (c)) can influence the generation of an action in two ways: the decision to act and the readiness to act. Different emotions correspond to different patterns of action.

3.5 *Neurotransmitters*

Modern brain science has disclosed that the brain functions of sensing, conception, and consciousness are controlled by brain chemicals called neurotransmitters. Ancient people had never imagined existence of such chemicals. Physiological data and information of our body state are communicated by the brain chemicals throughout our brain and body. They relay signals between nerve cells called neurons. Neurotransmitters affect breathing, heartbeat, mood, sleep, concentration, etc., and cause adverse symptoms when they are out of balance.

Neurotransmitters (NTs) are chemical messengers that transmit signals across a *synaptic cleft* from an *axon terminal* of one neuron to a *dendrite* of another "target" neuron (Fig. 3.6). Neurotransmitters are released from synaptic vesicles in an axon terminal into the synaptic cleft, where they are received by the receptors on the target cells.

There are two kinds of NTs: *inhibitory* and *excitatory*. Excitatory NTs are not necessarily exciting, but they are what stimulate the brain. The NTs that calm the brain and help create balance are called inhibitory. Inhibitory neurotransmitters balance mood and are easily depleted when the excitatory neurotransmitters are overactive.

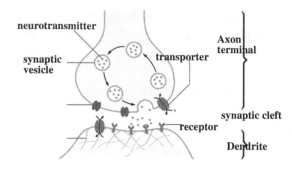

Fig. 3.6   Synapse (Wikipedia).

Dopamine and *noradrenaline* are two of the excitatory NTs, while serotonin is an inhibitory NT. All of them are monoamine (smaller) NTs and are described in detail below. *Endorphine* is a type of inhibitory NT, and it is involved in reducing pain and making a person calm and happy, and also responsible for the hibernation of bears. Another NT is *glutamate* (amino acid), an excitatory NT. Glutamate is involved in cognitive functions, such as learning and memory, in the brain.

Inside the brain, dopamine plays important roles in executive functions, motor control, motivation, arousal, reinforcement, and reward.

(i) ***Dopamine:*** Dopamine and noradrenaline are two of the excitatory NTs. Dopamine is a monoamine chemical. There are several distinct dopamine pathways, one of which plays a major role in reward-motivated behavior. Anticipation of most types of rewards increases the level of dopamine in the brain. Other brain dopamine pathways are involved in motor control and in controlling the release of various hormones.

These pathways form a dopamine system as a neuromodulatory. Dopamine pathways (Fig. 3.7) are neural ways in the brain that transmit dopamine from one region of the brain to another. Among them, the *mesolimbic* pathway ① transmits dopamine from the VTA ② (ventral tegmental area) to the *nucleus accumbens* ③. The VTA is located in the midbrain and identified as the origin of the dopamine cell bodies and as natural *reward circuitry* of the brain, solving problems of hunger or thirst.

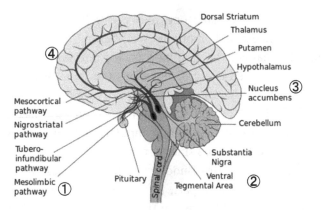

Fig. 3.7 Dopamine pathways.

It is important in cognition, motivation, orgasm, drug addiction, and intense emotions relating to love. The VTA contains neurons that project to the prefrontal cortex ④ (see Figs. 3.7 and 3.8).

(ii) *Noradrenaline* (also called norepinephrine): The general function of noradrenaline is to mobilize the brain and body for action. Noradrenaline release is lowest during sleep, rises during wakefulness, and reaches much higher levels during situations of stress or danger, in the so-called *fight-or-flight* response. In the brain, noradrenaline increases arousal and alertness, promotes vigilance, enhances formation and retrieval of memory, and focuses attention. It also increases restlessness and anxiety. In the rest of the body, noradrenaline increases heart rate and blood pressure. Noradrenaline is a substance that acts to increase the force of skeletal muscle contraction.

(It is released predominantly from the ends of sympathetic nerve fibers.)

(iii) *Serotonin*: Serotonin and endorphin are two of the inhibitory NTs. Serotonin is thought to be a contributor to feelings of well-being and happiness. The brain effect influenced by serotonin includes the regulation of mood, appetite, sleep, memory and learning, sexual behavior, temperature regulation, and social behavior. Dopamine, described previously, is important for attention, motivation, and goal-directed behavior and acts to reinforce behaviors that make one feel good. In contrast, serotonin is important for calmness and emotional well-being.

Regular physical activity is important for normal brain function and mental health. Sustained exercise increases the level of serotonin in the brain, and serotonin plays a key role in preserving brain function. Lower serotonin is linked to a greater impulsivity and an increased risk of suicide. Enhancing serotonin may reduce impulsiveness, and serotonin therapies are widely used to reduce suicide risk.

The traditional practice of Buddhist meditation goes back 2500 years. It was originally designed, in part, to relieve suffering. The practice has been successfully adapted to prevent the relapse of depression in modern therapy. The following is a report of recent neurobiological study [14].

The biological factors associated with mindfulness meditation have been studied on the basis of scientific approach. There are transformations in brain function during meditation, which have been documented using a variety of different methods. In general, results show increased signals in brain regions related to affect regulation and attentional control, with increased release of dopamine. Long-term brain changes are of greater interest to the mindfulness-based cognitive therapy as a preventive strategy. Several studies have compared brain morphology of experienced meditators, and findings include increased cortical thickness of the brain along with reduced age-related cortical thinning.

The neurons of the *Raphe nuclei* (Fig. 3.8) are the principal source of serotonin release in the brain, located along the midline of the brainstem. Axons from the neurons there form a neurotransmitter (NT) system reaching almost every part of the nervous system.

Activity of the serotonin nervous system is enhanced by (i) rhythmic exercises such as walking, jogging, dancing, or cycling; (ii) getting sunshine in the morning activates the serotonin nervous system. This is evidenced by the refreshing feeling when windows are opened in the morning. Sun-bathing longer than 30 minutes may inhibit the release of serotonin; (iii) Grooming behavior activates the serotonin release. Because of the widespread circuitry, activation of the serotonin system affects large areas of the brain. Elevated levels of serotonin result in increased feelings of empathy and closeness as well as a generalized state of well-being.

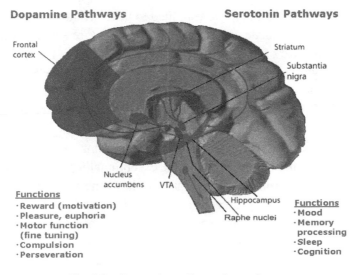

Dopamine Pathways

Serotonin Pathways

Frontal cortex

Striatum

Substantia nigra

Nucleus accumbens VTA

Hippocampus

Raphe nuclei

<u>Functions</u>
· Reward (motivation)
· Pleasure, euphoria
· Motor function
  (fine tuning)
· Compulsion
· Perseveration

<u>Functions</u>
· Mood
· Memory
  processing
· Sleep
· Cognition

Fig. 3.8 *Serotonin vs. Dopamine* pathways.

(iv) ***Endorphin*** (compared with dopamine): The name endorphin consists of two parts: *endo* and *orphin*; these are short forms of the words *endogenous* and *morphine*, intended to mean "a morphine-like substance originating from within the body." The principal function of endorphins is to inhibit the transmission of pain signals; they may also produce a feeling of euphoria, very similar to that produced by other opioids.

Endorphin and dopamine are two different kinds of neurotransmitters (NTs). Dopamine is called a monoamine NT. This means that it is a fairly small molecule. It is also a neuromodulator, which is explained as follows. Some NTs make a neuron more likely to fire, and other NTs make a neuron less likely to fire. Dopamine belongs to another category where it modulates other NTs' behavior.

On the other hand, endorphins are a whole class of NTs. It is not just one kind of molecule (not like dopamine), but a variety of molecules with similar shapes. Endorphins are larger molecules called peptides related in structure to proteins.

Sam Moss (2014) explained dopamine function by an example. Dopamine tends to be involved in reinforcing behavior. If one does something that one's brain is programmed to like, such as one eats a piece of their favorite chocolate at a bench of a liked park repeatedly, then one gets a small release of dopamine when one gets near the place. This is a signal that is going to subtly influence one's behavior in the future, causing to spend more time and buy a chocolate when one is there.

On the other hand, one of the endorphins' functions is numbing pain signals. Let us consider an example. When one's body is painful under a stress, the physiological body can produce endorphins so that one can still continue under what would otherwise be exceptionally painful conditions. Many long-distance runners claim that after they run for a while, they start to feel exceptionally happy, a condition sometimes called a runner's high. High levels of endorphins in response to the strain of running seem to be responsible for this state of mind.

The most important thing with these chemicals is that there is no inherent property in the chemicals themselves but rather in the brain circuits that they are involved in. The activity of these circuits is influenced by a huge number of factors and interactions between lots of different NTs.

(v) *Glutamate*: In our taste buds, the glutamate plays a role as a gustatory stimulus. Its receptors are thought to be responsible for the transduction of the *umami* taste stimuli.

Glutamate, also known as glutamic acid, is an amino acid and is the most common neurotransmitter in the body. Eighty percent of the brain's neurons release glutamate. Glutamate's most vital function as a neurotransmitter is in cognitive activities such as memory and learning.

Glutamate is involved in arousal, and anesthetic drugs seem to work at least partly by reducing neurotransmission normally regulated by glutamate. Glutamate is the most abundant neurotransmitter in our brain and central nervous system (CNS). It is involved in virtually every major excitatory brain function. While excitatory has a very specific meaning in neuroscience, in general terms, an excitatory neurotransmitter increases the likelihood that the neuron it acts upon will have an action potential (also called a nerve impulse). When an action potential occurs, the nerve is said to fire, where the firing of the neuron, in this case, is somewhat akin to the completion of an electric circuit that occurs when a light switch is turned on. The result of neurons firing is that a message can be spread throughout the neural circuit. It is estimated that well over half of all synapses in the brain release glutamate, making it the dominant neurotransmitter used for neural circuit communication (https://neurohacker.com/what-is-glutamate).

(vi) **Oxytocin**: Oxytocin ("quick birth," in Greek) is a mammalian hormone that also acts as a neurotransmitter in the brain. Oxytocin evokes feelings of contentment, reductions in anxiety, and feelings of calmness and security when in the company of one's mate.

Oxytocin is understood to influence intergroup perception via the mirror neuron system (MNS). The mirror neurons are a special class of brain cells that fire not only when an individual performs an action, but also when the individual observes someone else make the same movement. It is believed that we understand others not by thinking, but by feeling. Mirror neurons appear to let us "simulate" not just other people's actions, but also the intentions and emotions behind those actions. When a person A sees a person B smile (for example), A's mirror neurons for smiling fire up too, creating a sensation in A's mind of the feeling associated with B's smiling. Without thinking about what the person B intends by smiling, the person A experiences the meaning immediately and effortlessly. In this way, mirror neurons may explain how we can experience "second-hand" pain or emotion. Thus, by understanding the perspectives of other people, one responds with empathic concern.

Human empathy depends on the ability to share the emotions of others and to feel what other people feel. It is likely that the oxytocin signaling pathway works at the neurobiological level in the brain for enhancing individual's empathic feeling and developing altruistic action.

Oxytocin is synthesized by the neurons located in the hypothalamus. Oxytocin influences various social behaviors in addition to cardiovascular regulation and stimulation of uterine contractions during parturition and milk release during lactation. A new experiment pinpointed specific regions of the premotor cortex (PMC) that permit humans to understand and imitate such mirroring movements [14: G. Dewar 2016].

Chapter 4

Twelve Links of Dependent Origination

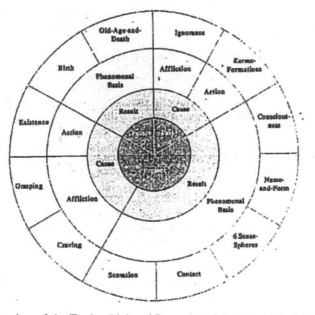

Diagrammatic view of the Twelve Links of Dependent Origination (clockwise from 12) ① *Inner most*; Past, Present, Future; ② Second: Cause, Result, Cause, Result; ③ Third: Affliction, Action, Phenomenal Basis, Affliction, Action, Phenomenal Basis; ④ *Outer most*: Ignorance, Karma-Formations, Consciousness, Name-and-Form, 6-Sense-Spheres, Contact, Sensation, Craving, Grasping, Existence, Birth, Old-age-and-Death.

4.1 General Introduction of the *Abhidharma* Philosophy

Buddha's Omniscience is a certain stream, just as fire, by the capacity of its flaming stream, thought to be able to burn all things one after another.

According to the *Abhidharmakośa* philosophy [2], the word "Buddha" designates a certain *stream* where "there is knowledge of all things one after another" and to this stream belongs the unique ability to immediately obtain accurate knowledge of the object in regard to anything one wishes to know [2, Chap. IX (p. 2543)]. Most importantly, note that one moment of thought is not capable of providing knowledge of everything.

(a) *Dependent origination*

The five aggregates are the dependent origination, and when viewed as a stream with twelve members, they are known as the twelve links of interdependent origination (twelve *nidāna*s):

1. Ignorance (*avidyā* 無明); 2. Karma formation (*saṃskāra* 行); 3. Consciousness (*vijñāna* 識); 4. Name-and-form (*nāma-rūpa* 名色); 5. Six sense-bases (*ṣaḍāyatana* 六处); 6. Contact (*sparśa* 触); 7. Sensation (*vedanā* 受); 8. Craving (*tṛṣṇā* 愛); 9. Grasping (*upādāna* 取); 10. Existence (*bhava* 有); 11. Birth (*jāti* 生); 12. Aging and death (*jarā-maraṇa* 老死).

The classic text of Master Vasubandhu [2] interpretes how the twelve links are distributed among the three existences, i.e. the previous existence (past), present existence, and subsequent existence (future). (i) *Ignorance* and (*karma*) *formation* are classified as being in the previous (past) existence, (ii) the subsequent eight links (from *consciousness* to *existence*) as being in the present existence, and (iii) the last two links (*birth* and *aging and death*) as being in the subsequent (future) existence.

The intrinsic nature of dependent origination with the above twelve factors is threefold, categorized as *defilement* (*kleśa*), *action* (karman), and *basis of phenomena* (*vastu*). (i) Three of the factors (i.e. ignorance, craving, and grasping) are *defilements* by their intrinsic nature. (ii) Two factors (i.e. karma formation and existence) are *actions*. (iii) Seven factors (i.e. consciousness, name and form, six sense-bases, contact, sensation, birth, and aging and death) are intrinsically the *basis of phenomena*, so qualified because they are the source of the defilements and actions. The factors comprising the phenomenal basis are effects, while the factors comprising defilements and actions are causes by their intrinsic nature.

(b) *Aggregates viewed as a stream*

It should be noted that the aggregates are viewed as a stream in the above discussion. In fact, the aggregates and the twelve links of dependent origination are represented as phenomena changing or evolving with respect to time. They describe features of the evolving world that are consistent with observations of modern science. To put it another way, existence is changing continuously. This is explained as follows: there exist momentary factors (dharma) in a stream that is similar to the stream of rice grains, as noted in the *Abhidharmakośa* [2, Chap. III, p. 958]. People transport rice grains to a distant village by passing through all intervening villages. In the same way, the mental stream takes birth from the previous moment with no discontiuity.

(c) *No-self (anātman)*

Non-Buddhists who believe in a self (*ātman*) will say: "If one admits that a sentient being goes to another world (according to the reincarnation concept), then the self in which we believe is established." The Master says that the self does not exist. The self in which they believe is, by incarnation, an entity that abandons the aggregates of one existence and seizes the aggregates of another. This *self*, a soul (an internal agent), does not exist. The *self* is only a name given to the aggregate. The existence of a *self* as a *designation* (i.e. only a name that identifies an object) is not denied. Action (karma) exists. Its effect exists nominally as dependent origination. However, regardless of the causal relationship between phenomena (dharma), there is no agent that abandons previous aggregates and seizes new ones.

The Master is definitely not saying that the aggregates pass away or transmigrates into another existence. What is mentioned by the causal relationship is interpreted as follows: "When this exists, that comes to be. From the arising of this, that arises." This defines the concept of *dependent origination* (Sanskrit: *pratityasamutpada*). This is nothing but a scientific statement, or a scientific point of view.

The aggregates are transient, and the aggregates themselves are incapable of transmigration. In the absence of any *self* or any permanent entity, i.e. with *no-self*, the stream of aggregates is conditioned by the *dependent origination*. It is inevitably influenced and controlled by defilements and karmic actions, as will be explained later.

(d) *Defilement*

Ignorance (*nescience*), *craving*, and *grasping* are all characteristics of defilement, which is an active state of the living mind. Its nature is explained to be like that of (1) a *seed*, (2) a *root*, (3) a *tree*, (4) a *husk of grain*, and (5) a *nāga* [2, Chap. III, p. 1030].

1. From the seed, there arises a sprout, stalk, leaves, etc. In the same way, from defilement there arises another defilement, action, and phenomenal basis.
2. A tree whose root is not cut off continues to grow even if its greenery is cut again and again. In the same way, as long as the defilement is not cut off, the state of mind continues to grow.
3. As a tree grows, it produces flowers and fruits at different times. In the same way, defilement grows to produce defilement, action, and the ripening of the fruit of the phenomenal basis, such as *consciousness*, *six sense-bases*, *contact*, *sensation*, and *birth*.
4. When a grain is stripped of its husk, it does not germinate properly. In the same way, in order to effectively shoot forth another birth, action must be associated with defilement (acting like a husk).
5. The pond where *nāga*s live does not dry up. In the same way, the ocean of existence where a defilement (a *nāga*) lives does not dry up.

(e) *Life-force*

Let us recall the five aggregates that were discussed in Chapter 2. The fourth aggregate, i.e. the aggregate of *formation*, is an action (karma) in nature. It creates and determines the five aggregates of the future states. It is like a river, i.e. a stream of water. The stream of aggregates continues to grow gradually, and in due course the action ripens as *life-force* (Sanskrit: *āyuhsamskara*). *Life* is distinguished from *death* by an assumed *vital faculty* (*jivita*) which is capable of supporting *warmth, consciousness* and *life-force*. These components distinguish *sentient* beings from *insentient* matter [2, Chap. II, p. 575 & its endnotes].

According to the philosophy of the twelve links of dependent origination (or transmigration), the stream of aggregates influenced by actions and defilements (characterized by *ignorance, craving,* and *grasping*) enters the mother's womb, and this stream is fertilized and prolonged by *embryogenesis,* and evolves as a stream which constitutes an intermediate existence between previous existence (before fertilization) and the subsequent one (after birth). In the intermediate state, the faculties mature and the defilements become active, from which actions arise. After birth and growth, and when the body perishes, the stream passes into another existence depending on these defilements and actions.

Birth depends on defilements and actions (see below); defilements and actions depend on birth. In this way, the wheel of existence is without beginning.

(f) *Beginningless wheel of existence*
All dharmas arise because of causes and conditions. The world as a whole does not arise from a single cause that one calls God, Self, or by any other name. This is interpreted by the *Abhidharmakośa* [2, Chap. II, p. 675] as follows. If things were produced by a single cause, they would arise all at the same time. But everyone sees that they arise successively. They depend on causes that are themselves dependent on other causes. This would imply that the stream of causes has no beginning. In order for the wheel of existence to begin, it would be necessary that the beginning part of the wheel should not have causes. But we see that a seed produces a shoot, or that from an egg a bird is born, and vice versa. There is no birth or appearance that does not have causes. If one dharma arises without cause, then all dharmas will arise without cause. Thus the cycle of existence is beginningless. If birth arises depending on causes, it will not arise if its causes are destroyed, just as a shoot will not arise if its seed is burned.

In the philosophy of the *twelve links of dependent origination,* the twelve members are distributed among three existences. In particular, *ignorance (nescience)* and *(karma) formation* are classified as being in the previous (past) existence. The subsequent eight members (from *consciousness* to *existence*) are in the present existence. The last two members are in the subsequent existence.

(g) *Abhidharmakośa*

Vasubandhu (c. 320–400 CE) composed the *Abhidharmakośa* (*AK*) in verse following the traditional style. He skillfully summarizes the *abhidharma* philosophy, occasionally inserting the word *kila* ("alleged") where he dissents from orthodoxy. In his own *prose* commentary, the *Abhidharmakośa-bhāṣya* (*AKB*), he reviews the orthodox tenets, often criticizing them from the perspective of the Sautrāntika school [2, p. 6, Commentary by Bhikkhu Dhammajoti]. The interpretation of Section 4.5 is taken from "Vasubandhu's brief presentation of the sequential causation of the twelve members of dependent origination" [2, Chap. III, pp. 1001–4].

4.2 Nescience (Ignorance, *Avidyā* 無明)

What is *nescience* (ignorance, *avidyā*)? One recalls that the word "Buddha" designates a certain stream referred to by the term "omniscient being (*Sarva-vidyā*)". *Nescience* is non-*vidyā* (*avidyā*, ignorance), which denotes lack of *clear knowledge* (*vidyā*). It is like a non-friend or enemy — the opposite of a friend. The Chinese translation of *avidyā* is 無明, interpreted as *no clarity*. This denotes the lack of understanding that what is produced through *dependence* (in other words, dependent origination) is *merely* "conditioning" (i.e. negating the *self*). This indicates the *fundamental* ignorance, i.e. ignorance that is independent of, not associated with, other factors such as *linked existence of interdependence*.

Ignorance is explained with a long list of different kinds of ignorance in the endnotes to the *conditioned factors* [2, Chap. III, p. 1002]. On replying to the question "Of what type is *ignorance*?" in the sutra *Pratītyasamutpāda-sūtra*, the answer is a stream of everything of *not-knowing*'s:

> Not knowing the past, not knowing the future, not knowing deeds, not knowing their maturation, not knowing the Buddha, not knowing the community, not knowing suffering, not knowing the origin, not knowing

cessation, not knowing the path, not knowing the causes, not knowing the factors arisen through the causes, not knowing wholesome and unwholesome, blameworthy and blameless, to be attended to and not to be attended to, low and high, and not knowing dependently originated factors. Not knowing what is in accordance with truth, non-seeing, darkness, delusion, obscurity, this is called *ignorance* or termed as nescience.

This implies that *nescience* is the *fundamental ignorance.*

As mentioned above, ignorance is the lack of clear knowledge, or non-clarity. Fools or ordinary worldlings are *intent on* (absorbed in) a defiled view of the *self* and driven to perform threefold actions, i.e. bodily, vocal, and mental, with the aim of acquiring pleasant sensations or neither unpleasant nor pleasant sensations for themselves. However, even wise men (or wise women) are driven to perform three types of defiled actions: bodily action, vocal action, and mental action. In fact, living in the beginningless cycle of existence, how are they able to be omniscient? Human beings are not sure about not only the long history of biological evolution, geological ages, or cosmological evolution, but also about their own ancestors or homes of generations before. They have no memory of when they were in their mothers' wombs. How it is reasonable to say that every human being is well versed in everything, and hence has clear knowledge and behaves appropriately? Moreover, there is a fundamental issue as to whether one can control the internal state or activity of one's brain, whether one can control the state or activity of one's internal organs, or whether one can maintain the body to function in its best condition. These are everlasting challenges for Buddhists and human beings.

Ignorance is not defiled understanding because ignorance is defined as defilement of understanding; hence it is not "understanding." The Sanskrit term for it, *avidyā*, is defined as the cause or condition of karma formation in the twelve links; therefore, ignorance is a real separate dharma, which is explained in [2] as a *bondage*, a *proclivity* (a root of existence), a *flow* (*āsrava*, interpreted as an influential sense impelling the soul towards external objects), and a *rapid stream* (*ogha*) [2, Chap. III, p. 1005].

In other words, ignorance means the stream of five aggregates that is in the state of defilement. Ignorance goes with defilements. All defilements, in fact, accompany ignorance and become active through ignorance. It is said that ignorance is like a king. If the king is coming, his courtiers (defilements) accompany him.

4.3   From Previous Existence to Present Existence (Twelve Links — I)

In the twelve links of dependent origination, the second member, *karma formation* (*saṃskāra*, 行), refers to the action formations that influence the third, *consciousness*. In the general sense, *saṃskāra* refers to the power to form things, and the *saṃskāras* are all the action (karma) formations driven by *ignorance*. The action is *intention*, or that which is produced through intention. The intention is mental action, from which two actions arise: bodily action and vocal action. *Saṃskāra* refers to forces that form everything in the world. Action (karma) formations exist in dependence on ignorance, and the past action (karma) influences consciousness. (The same *saṃskāra* refers to the fourth of the five aggregates: formations or, particularly, volition.)

A sequence of karma formations (*saṃskāra*) is regarded as the stream and the state of actions (which are meritorious or non-meritorious, etc.) of the five aggregates carried out in the previous existence. Consciousness (*vijñāna*, 識), the third member of the twelve links, is the five aggregates in the womb at conception. How consciousness arises due to karmic formations? Due to the effective force of previous action (and fertilization), the stream of consciousnesses (i.e. evolution of *embryo*) goes to such and such a destination (*remote* as it may be) where a life-flame is said to be alive. Consciousness is the entire stream of consciousnesses in the intermediate existence, i.e. in the womb. [2, Chap. III, p. 1002]

What is consciousness? According to modern scientific biology, the entire stream of consciousnesses corresponds to the embryogenesis evolving in the mother's womb. *Human embryogenesis* (stream of consciousnesses) is the process of cell division after fertilization and cellular differentiation of the embryo that occurs during the early stages of development. In biological terms, the development in the womb entails growth from a one-celled *zygote* to a human baby. This is the entire stream of consciousnesses.

Anyway, the scenario of the stream of twelve links of dependent origination is presented up to the third member, *consciousness*, according to the *Abhidharmakośa* [2, Chap. III, pp. 978–1004]. Here we interrupt the stream and present another scenario of the stream of twelve links of dependent origination because there are four types of dependent origination and the *Abhidharmakośa* stream is interpreted as "prolonged" or "pertaining to states" (with each comprising five aggregates). The remaining part of the *Abhidharmakośa* stream is postponed to Section 4.5.

4.4 Scenarios of Different Type of Dependent Origination

Dependent origination is fourfold:

a. *Momentary* or *instantaneous*: The whole process proceeds simultaneously and successively;
b. *Prolonged*: This refers to the cycling pertaining to the previous and subsequent periods of existence. The defilements, karma, and its retribution are not confined to just the immediately preceding and succeeding existences, but extends to beginningless past and boundless future existences;
c. *Connected*: Conjoined with the cause-effect connection;
d. *Pertaining to states*: Twelve states, each comprising five aggregates, continue to arise uninterruptedly, manifesting their efficacies as *dharmas*.

The scenario of the stream of dependent origination of the previous section is the types *b* and *d*.

Let us consider a sequence of twelve links of *momentary* dependent origination of the class *a* by taking particular examples of twelve links of human behavior such as committing murder (or falling in love). The twelve members are as follows:

1. *Ignorance* is the defiled proclivity for violence (or their inconstant emotion);
2. *Karma formation* is caused by the intention of hatred (or of affection);
3. *Consciousness* is the distinct delusive perception of particular target object-beings.
4. *Name-and-form* is the aggregate (form, sensing, conception, formation) that has co-arisen with the *consciousness* of violence (or emotion);
5. *Six sense-bases* are the sense faculties;
6. *Contact* is the functioning of the six sense-bases;
7. *Sensation* is the experience of contact;
8. *Craving* is greed (or love);
9. *Grasping* is hostile sieging (or outpouring of emotion);
10. *Existence* is bodily or vocal actions that proceed from craving and/or grasping;
11. *Birth* is *the* emergence of all factors resulting from ill deeds (or affairs);
12. *Aging and death* is the maturity and breaking up of these factors.

In Section 4.6, we will see another example of type *b+d* of the twelve links of dependent origination.

4.5 From Present Existence to Subsequent Existence (Twelve Links — II)

Let us continue with the stream of dependent origination (from Section 4.3) from the present existence to the subsequent existence, according to *abhidharma* and Vasubandhu [2, Chap. III, pp. 978–1004].

Next to *consciousness* (the third member) as an antecedent, there arises *name-and-form* (the fourth stage). What is form? It denotes all material forms. What is name? It denotes the (four) immaterial aggregates (other than form). These two are called *name-and-form*. However, at the stage of *embryo*, only the body's sense faculty exists, whereas other sense faculties (such as eye or ear) are not yet manifested (name-and-form).

It is at the moment when the sense faculties mature sufficiently and begin to function that the pre-existing sense faculty merges with other sense faculties to become the complete the full set of six sense faculties. This is called the *six sense-bases* or *sense spheres* (the *fifth* stage), which are the five aggregates from the first manifestation of the six sense-bases until the next, sixth stage: *contact.*

When the sense faculties encounter their object, consciousness arises. Through the coming together of the three, i.e. the *sense faculty*, its *object field*, and *consciousness*, there is *contact*. There are six groups of contact: contact with the eye or otherwise, which can be experienced as pleasant or unpleasant. The stage of contact lasts until the contact becomes capable of distinguishing sensations such as "This is a cause of pleasure."

From contact, threefold *sensation* (the seventh stage) arises: *pleasure, displeasure,* and *neither pleasure nor displeasure.* Five groups of sensations arise from the contact of the bodily sense faculties with their basis (*āsraya*), such as the eye. The sixth sensation arises from the contact of the mental faculty (*manas*). Its basis is the brain within the skull from the point of view of modern brain science.

From sensation, there arises the eighth, *craving* (such as desire, greed, covetousness, love, and passion). Craving (*tṛṣṇā*) is the cause for the production of existence: just as the seed is the cause for a sprout, such as the rice sprout or the sprout of two embryonic leaves of a dicots plant. A stream of thoughts bends unceasingly toward the object with regard to which one has craving. There is no defilement as much as craving that adheres to the person [2, Chap. VI, p. 1891].

From *craving* in relation to the desired sensation, there arises fourfold *grasping* (the ninth): (i) grasping of desire; (ii) grasping of view; (iii) grasping of morality and spiritual practices; (iv) grasping of assertion of the self. Grasping of morality usually leads to the rejection of immorality.

Assertion of the self is the afflicted view of "I am this or that" and the concept of "I am this or that." One asserts the *self* as a result of those two, with which one mentally formulates the notion of a self (*ātman*). It is *assertion* because the self itself does not exist as a real entity. It is called grasping of the *assertion* of the self because of the *non-existence* of the self and because of the *grasping at* a *mere designation*. These two reasons are in fact a declaration of the *emptiness* of self. It is said in a sutra:

> *Fools who are not learned and the ordinary worldlings who fell into the mere verbal notions, think "I am" or think "mine". But there is no self nor mine.*

Grasping of desire, afflicted views, etc., is predilection and attachment to them.

In response to craving or grasping (running around in this manner), bodily or vocal actions proceed and produce a new existence in the future. This is the tenth, *existence* (*bhava*), which is the state of five aggregates seeking enjoyment or avoiding risk. There are three types: existence in the sphere of desire, that of the material sphere, and that of the immaterial sphere.

Depending on the existence, there is *birth* (eleventh), the emergence of all factors, i.e. the future life by means of the entry of consciousness. Being name-and-form in its nature and becoming visible, birth entails the five aggregates.

Because of birth, there is *old age* and *death*, which is the twelfth stage of the twelve streams of the five aggregates. With regard to this future existence, four members (name-and-form, six sense-bases, contact, and sensation) are designated by old age and death. The maturity of these factors is old age (*jarā*) because of diminished capability, and the breaking up of these factors is death (*maraṇa*). The fading away of the sense organs is old age. The fading away of life heat and disintegration of the sense organs is called *death*.

In this manner, there arises the entire and mere mass of *unsatisfactoriness*, i.e. an accumulation of suffering that has neither beginning nor end. The term *mere* denotes "without any relation to any self."

4.6 Abhidharma Philosophy

Vasubandhu (c. 320–400 CE) composed the *Abhidharmakośa* (AK) in verse following the traditional style. He skillfully summarized the *abhidharma* philosophy, occasionally inserting the word *kila* ("alleged"), where he dissents from orthodoxy. In his own prose commentary, the *Abhidharmakośa-bhāṣya* (*AKB*), he reviews the orthodox tenets, often criticizing them from the perspective of the Sautrāntika school [2, p. 6, Commentary by Bhikkhu Dhammajoti].

According to the *abhidharma* philosophy, the twelve members of dependent origination are categorized into three types based on their intrinsic nature: *defilement, action* (karman), and the *basis of phenomena.* (i) *Defilement* is three members: ignorance, craving, and grasping; (ii) *action* is two: karma formation and existence; (iii) the *basis of phenomena* is seven members: consciousness, name-and-form, six sense-bases, contact, sensation, birth, and aging and death.

From the five aggregates of *sensation* (i.e. *sensing,* a phenomenal basis), there arise the defilements of *craving* and *grasping.* According to the *abhidharma* philosophy, verses 23 and 24 of Chapter III are as follows:

23. *Sensation is the five aggregates before the timing of sexual union (i.e. when attachment to sexual union is not in action). Craving is the state of five aggregates of those who desire enjoyment and sexual union. Grasping is to be distinguished from Craving. It is the state of five aggregates of those who run around in search of the enjoyment.*
24. *Running around in this manner, Existence is the state of five aggregates when one performs actions (karma) which will have for their effect a future existence. Birth is the state of five aggregates of reincarnation. Old-age-and-death is the state of five aggregates until Sensation.*

The last, *old age and death,* designates the four members of future existence: name-and-form, six sense-bases, contact, and sensation. As it will arise, *existence* is a synonym for *karma,* causing one to migrate to the next existence. When examined closely, the above is nothing but the links of a sequence of dependent origination.

Of the twelve links (see the circular diagram of the first page of Chap. 4), *ignorance* is the cause of the current birth, and *craving* and *grasping* are the cause of future birth. The links of *formation* and *existence* are *karma,* causing the births of the present and the future. The seven links of consciousness, name-and-form, six sense-bases, contact, sensation, birth, and aging and death are all bases, because they are the support of defilements and karma. The first fives are the results of *ignorance* and *formation*, and the last two are the results of *craving*, *grasping*, and *existence*. The present lifetime is explained in detail by five results, while the past life and future life are explained concisely by two causes and two results, respectively. This is because one can infer them from the detailed explanation of cause and effect of the middle life.

Chapter 5

Four Noble Truths

Buddha in Sarnath Museum
(Dhammajak Mutra) [15]

5.1 First Turning of the Dharma Wheel at the Deer Park

When Gautama Siddhārtha was only nine years old, he sat in meditation beneath the cool shade of a rose-apple tree. The memory of this experience was imprinted in his mind for the rest of his life.

According to the Buddhist sutra (*Dhamma-cakkappavattana Sutta, The Discourse That Sets Turning the Wheel of Truth*), Gautama Buddha addressed the Four Noble Truths first after he had been substantially absorbed in enlightenment in Bodhgaya. In his address, there were four key verses representing the Four Noble Truths. In the subsequent sections of this chapter, the philosophical aspects of the Four Noble Truths are presented according to the *Abhidharmakośa* [2, *AKB*] of Vasubandhu. Before presenting them, it would be interesting to see a story of the First Sermon after the Bodhgaya enlightenment, by citing a passage from the text of Master Thich Nhat Hahn [6] (Chap. 22). Through his awakening, Gautama Siddhārtha found what Buddhists call the Middle Way — a path of moderation between the two extremes of indulgence and self-mortification. Gautama realized that one cannot make a fire with soft, wet wood. He realized that body and mind form one reality and that they are not separate.

Gautama Siddhārtha then traveled to the Deer Park near Varanasi. When he entered the gate of the Deer Park, he found five ascetic companions with whom he had practiced at the caves of the Dungeshwari mountain before descending to Uruvela village and then to Bodhgaya. The story of the text [6] is as follows.

The five former companions now practicing there were so impressed by his radiant bearing that they all stood up at once. Siddhārtha seemed to be surrounded by an aura of light. Each step he took revealed a deep spiritual strength. His penetrating gaze undermined the intention of the five ascetic companions (Kondanna, Mahanama, Baddhiya, Vappa, and Assaji) to snub him. Kondanna ran up to him and took his begging bowl. Mahanama fetched water so that Siddhārtha could wash his hands and feet. Baddhiya pulled up a stool for him to sit on. Vappa found a fan of palm leaves and began to fan him. Assaji stood to one side, not knowing what to do.

> After Gautama Siddhārtha washed his hands and feet, Assaji realized that he could fill a bowl with cool water and offer it to him. The five friends sat in a circle around Gautama, who looked kindly at them, "My brothers, I have found the Way, and I will show it to you."
>   ...   .
> The Buddha said, "Then please listen, my friends, I have found the Great Way, and I will show it to you. You will be the first to hear my Teaching, This Dharma is not the result of thinking. It is the fruit of direct experience. Listen serenely with all your awareness."

... ...

"Brothers, there are four truths: the existence of suffering, the cause of suffering, the cessation of suffering, and the path which leads to the cessation of suffering. I call these the Four Noble Truths."

5.2 How Suffering is this World?

In the *Abhidharmakośa* (*AKB*) [2], the author Vasubandhu begins from the first verse:

> *The Buddha Bhagavat* (Glorious Buddha) *has destroyed all darkness in every respect.*
> *He has drawn out the world from the mire of Samsara* (transmigration). *I pay homage to him, to the teacher of truth, before propounding the treatise Abhidharmakośa.*

Darkness is also called *nescience* (*avidyā*, ignorance), because ignorance (nescience) prevents one from seeing things as they are. *Avidyā* was the fundamental obstacle of worldly existence which Buddha destroyed. Worldly existence is also known as *saṃsāra* (transmigration), in which *sentient beings wander about and migrate back and forth between successive existences of birth and death.* Such a world is called the *mundane* world, which indeed is a mire, because the world remains stuck in the mire and attached to it. Human beings are mired in the mundane world. It is difficult to cross over such existences under the darkness of nescience. People are suffering existences in the mundane world.

There are a *variety of existences* of sentient beings in the mundane world that arise from the diversity of their actions in the world. From a scientific and biological point of view concerning our living body and brain, there exists an inherent system of physiological mechanisms within human bodies which transform us internally and impel us (mind and body) to external objects. This is an internal transformation and a biological function of the innate brain/body possessed by human beings. It is a latent defilement from which they are unable to get away. Internal physiological mechanisms are purely scientific processes that are not fundamentally defiled themselves (consistent with *nirvāṇa*). But there exists a mental factor in human beings, and internal transformation is accompanied by a latent defilement (a mental factor). The latent defilement, however, is susceptible to abandonment through cultivation practices, as discussed in more detail below. In fact, the internal physiological mechanism transforms human beings in their journey to awakening. The main concern of the abhidharma philosophy is how to achieve it systematically. How a human is able to abandon defilement?

The concept corresponding to the latent defilement is described by the special word *anusaya* or *āsrava* in Sanskrit. The *anusaya* corresponding to the latent defilement is translated as proclivity in English, which represents defilements characterized as attachment to sensual *desire, attachment to pleasure, or to ignorance, hostility, conceit, afflicted view*, or *doubt* [2, Chap. 5]. Any of these proclivities causes one to be *deluded* with respect to an external object of consciousness. Because defilements (or proclivities) are subtle, it is difficult to detect their emergence. However, a proclivity binds with the mental stream to become a continuous serial flow. It is difficult to disconnect both of them (latent proclivity and physiological mental stream). This is an inclination arising from nescience (*avidyā*), which causes suffering. By means of the proclivity, the mental stream flows out into the object field (external objects), driving one to approach the object field. Thus, the proclivity is called also flux or discharge (*āsrava*). The same proclivity is also called grasping (*upādāna*) because, due to it, one tries to grasp enjoyable things. Eventually, one gets suffering. Under ignorant nescience, one is constantly followed along by such a physiological mechanism of the serial flow. It is beginningless. This is the background of the first of Four Noble Truths.

The proclivity described by *anusaya* is a synonym of defilement (*kleśa*) and used to signify its subtle and tenacious nature. Of all the proclivities, nescience (*avidyā*) is the root. However, the proclivity of *attachment, hostility,* and *conceit* is susceptible to being abandoned by cultivation practices, just like the elephant keeper, being on the elephant, tames the elephant. The woodcutter climbs on the tree and cuts it. The essence of the Noble Truths is the statement that there exists the supreme path that brings about the abandonment of ordinary worldly status (defiled mental stream). The Noble Truths to be given below state implicitly that there exists a fundamental stream of vital life free from defilements. The proclivity tenaciously attached to it dies down gradually through the practice of the Path. This is the third truth of *cessation* (Section 5.7) and the fourth truth of the Path (Section 5.8).

The proclivity is subtle, making it difficult to recognize before the defilements manifest themselves. Citing an example, Venerable Ananda says: "I do not know if I produce or if I do not produce conceit toward my companions." This does not imply that he does not produce conceit because the mode of existence of the proclivity "conceit" is difficult to recognize. The same applies for the other proclivities.

The cause (origin) of suffering is craving or mental thirst. One continually searches for something outside oneself to satisfy oneself or to make oneself happy. But no matter how successful, one never remains satisfied because craving grows from ignorance (nescience). A mental factor (a latent defilement) then drives personal actions. One goes through life grabbing one thing after another in order to get a sense of satisfaction. One attaches oneself not only to things but also to ideas about oneself and the world. Then one becomes frustrated when the world does not behave the way one thinks it should and one's life does not conform to its expectations. Eventually, suffering results. This is the conceptual basis of the first two of the Four Noble Truths. In any case, the proclivity is eventually calmed (according to the third and fourth of the Noble Truths).

Secondary defilements arise following the main defilements. There are two types. One (called *envelopment*) is characterized by *non-modesty, shamelessness, jealousy, greed, restlessness, regret, torpor, sleepiness, anger*, and *concealment*. These defilements bind sentient beings and place them in the prison of saṃsāra; hence they are called *envelopments*. These serve as the causes of various evil actions.

One experiences *unsatisfactoriness* that arises from the envelopments, which are the outflow of the defilements caused by attachment to self-pleasure.

The second one (called *defilement stains*) is characterized by *deceit, pride, harmfulness, enmity, opinionatedness*, and *dissimulation*. These stains emanate from defilements, just as sweat or pus oozes out of the body, and harm or delude others. Attachment leads to deceit and pride.

Hostility gives way to harmfulness and enmity. When bad views are given precedence, there arises depraved opinionatedness, and afflicted views lead to dissimulation [*AKB*, 2, Chap. 5].

5.3 Four Noble Truths (Four Ārya Satyānis)

What is the Noble Truth (*ārya-satya*)? It is the truth that is true for the noble ones (*ārya*). The truth of the noble ones is the Noble Truth. The noble ones (*ārya*) are those who *have crossed over to the other side* on account of their perfection in disconnection from defilements. What is perfection? It is the complete abandonment of *attachment*, of *hatred*, of *delusion*, of all *defilements*. That is to say, perfection is *nirvāṇa* [*AKB*, 2, Chap. 3]. The noble ones denote the Buddhas, the *pratyeka*s (*lone buddhas*), and *śrāvaka*s (hearers). From a wider point of view, the noble ones are *bhikhṣu*s and *bhikhuṣuni*s who are ordained Buddhists, and *upāsaka*s and *upāsika*s who are lay devotees. The noble ones see the Truths as they are. The seeing of others (ordinary worldlings) is erroneous.

In Section 2.4 (Five Aggregates), all *dharma*s (phenomena) are explained as composite (*saṃskṛta*, 有為) or noncomposite (*asaṃskṛta*, 無為). "Composite" means *those formed by causes and conditions coming together and coexisting in assemblage*. In this sense, the composite is said to be *conditioned*. The five aggregates are *forms, sensations, conceptions, formations*, and *consciousnesses*, which all are conditioned and said to be *impure* if the proclivity of defilement adheres to them or grows in them. The five aggregates are conditioned. Therefore, they are impure, with the exception of those that are part of the pure, noble path.

Worldly existence is fundamentally unsatisfactory. There is a way out of this suffering existence. It is like having found a disease. There follows the search for its cause, treatment, and elimination [*AKB*, 2, Chap. 6, p. 1877]. Thus, the Four Noble Truths are as follows:

(i) Practitioners first examine how one is attached, how one is oppressed, and in fact how one is suffering. One realizes the suffering and seeks to be liberated from it. This is *the truth of suffering* (duhkha).

(ii) Next, the practitioners ask, what is the cause? And one examines *the truth of origin*.

(iii) Next, they ask, what does cessation consist of? And one examines *the truth of cessation.*

(iv) Next, they ask, what is the path to cessation? And one examines *the truth of the path.*

The First Noble Truth is *duḥkha* (suffering) of the mundane world. All of the Four Noble Truths are:

(i) The truth of *duḥkha* of the mundane world.
(ii) The truth of the origin of *duḥkha* (*samudāya*);
(iii) The truth of the cessation of *duḥkha* (*nirodha*);
(iv) The truth of the Path (*mārga*).

5.4 The First Noble Truth: *Duḥkha* (Suffering) of the Mundane World

The concept of suffering (*duḥkha*) plays a central role in the motivations of Buddhists. It is said in an old text: "Birth is unsatisfactory; old age, disease, death, etc., are unsatisfactory. That is, all of this is the ground for suffering." The Sanskrit *duḥkha* is used everywhere where unsatisfactoriness is derived from *impermanence*. According to academic studies of ancient Indian philosophy, there are three types of unsatisfactoriness [2, Chap. 6]: (i) unsatisfactoriness of painful sensation; (ii) unsatisfactoriness upon decay. When a pleasant sensation subsides, it is replaced by displeasure. As a matter of fact, all sensations are subject to change and decay over time, which is why they are *impermanent* and *unsatisfactory*; (iii) the *unsatisfactoriness* attributed to all conditioned factors of existence due to their *impermanence*. However, one must realize the universal *unsatisfactoriness* of existence as a result of the conditioned dharmas of existence being distributed among the above three types of unsatisfactoriness. This is known as the *universality of suffering* in Buddhist philosophy.

Duḥkha denotes not only *unsatisfactoriness* but also *pain, suffering, sorrow*, or *unpleasantness*. In Buddhist philosophy, *impure composite things* in the mundane world are all a cause for suffering and unsatisfactory in the sense of *duḥkha*. All of them are conditioned existence and are subject to change and decay over time, and hence *impermanent*. Thus, all impure (i.e. defiled by proclivities) composite dharmas in the mundane world lead to suffering and unsatisfactoriness because of their impermanence. Thus, we have

The First Noble Truth: duḥkha of the mundane world

The Sanskrit *duḥkha* is usually translated in English as *suffering*, *painfulness*, or *unsatisfactoriness*. If one of them is chosen, it may be misleading since other meanings might be overlooked. It refers to the universal *unsatisfactoriness*, painfulness, and suffering of mundane life. *Duḥkha* also includes the *sorrow* or *grief* at parting or when one of our relatives or friends passes away. Thus, it would be better to keep the original Sanskrit *duḥkha* in order to represent the central concept described above in Buddhist philosophy.[1] Hence, all impure composite things are *duḥkha* in the *fundamental* sense. Composite things have no intrinsic nature (i.e. they are devoid of intrinsic nature; see Chapter 9 of this book). Thus, the *duḥkha* of the mundane world constitutes the First Noble Truth, which inspires the subsequent three Noble Truths, leading to *nirvāṇa*.

There are three kinds of sensation: *pleasant, unpleasant,* and *neither unpleasant nor pleasant.* Living in this world can be very pleasant or joyful at times. However, it is said in a sutra that *sensation*, whatever it may be, is included in *unsatisfactoriness.* An unpleasant sensation is unsatisfactory because it is painful in its intrinsic nature. A neither unpleasant nor pleasant sensation is unsatisfactory because of its impermanence. However, even a pleasant sensation is unsatisfactory because it is subject to change, and therefore impermanent. In fact, a pleasant sensation is pleasant when it arises, but unsatisfactory when it changes. The sensation is of a temporary or instantaneous nature, while the concept of satisfactoriness is concerned with *truths* prolonged temporally. All conditional phenomena and experiences are not ultimately satisfying.

---

[1] Hinted at in Barbara O'Brien (2015), The four noble truths (Foundation of Buddhism). http://buddhism.about.com/od/thefournobletruths/a/The-Four-Noble-Truths.htm

Duḥkha denotes *suffering, anxiety, stress, or unsatisfactoriness*. The Noble Truth of *duḥkha* is one of the most important concepts in Buddhism. The Buddha said:

I have taught one thing and one thing only, duḥkha and the cessation of duḥkha.

5.5 The Second Noble Truth: The *Origin of Duḥkha*

The five aggregates are all impure composite dharmas (i.e. defiled by proclivities) in the mundane world. Because of their impermanence, they are sorrowful and unsatisfactory. Therefore, the First Noble Truth (*duḥkha* of the mundane world) is an effect of (caused by) the five aggregates. *Duḥkha* originates from the five aggregates, and this is nothing but

The Second Noble Truth: the origin of duḥkha

The origin of *duḥkha* is in the five aggregates: *forms, sensations, conceptions, formations,* and *consciousnes,* which all are conditioned and impure.

There is no other defilement that adheres to a person as much as craving. Craving is said to be the cause of *duḥkha,* but all other impure factors are also the causes. The other factors are named in a sutra as follows. "Action (*karma*), craving (*tṛṣṇā*) and ignorance (*avidyā*) are the causes that produce the future conditioned composite things which are nothing but the *duḥkha* in the *fundamental* sense. Those will make up the stream of existence of the self, which is suffering." Craving is the origin of unsatisfactoriness due to its major importance.

As seen in the details, craving is the cause of re-existence, just as the wheat seed is the cause of a wheat sprout. Craving denotes a strong desire for something. Because the stream of thoughts bends through craving, the stream bends unceasingly toward the object that is craved. Hence, craving is something like a force following mechanical laws. It is the cause of future existence; in other words, it is the driving force that creates the future. In response to craving, bodily or vocal actions occur, resulting in the creation of a new existence in the future.

As a result of their effect on the mundane world, the five aggregates give rise to the *unsatisfactoriness* mentioned in the First Noble Truth. On the other hand, the same five aggregates are the Second Truth of origin because they are a cause. The First and Second Truths are truths of the noble ones, because non-noble others are not satisfied with what the noble ones call satisfactory [*AKB*, 2, Chap. 6, §ACA].

Noble Truth is called the truth of the noble ones, not the truth of non-noble others, because the seeing of the non-noble ones is erroneous. In fact,

> *with what the noble ones call satisfactory, non-noble others are not satisfied; that which non-noble others call satisfactory, the noble ones call unsatisfactory.*

Generally speaking, all the forces (like craving) making composites are unsatisfactory when seen by the noble ones only under the above aspect. Putting it figuratively,

> *When an eyelash is placed on the palm of the hand, one usually does not feel anything at all. But the same eyelash, getting into the eye, causes suffering and harm. Likewise, ordinary worldlings (corresponding to "the hand") do not perceive the eyelash as unsatisfactory because of conditioning. However, the noble ones (similar to the "eye") are tormented by it.*

The sections (after Section 5.6) present the Third and Fourth Noble Truths, which are truths of both the noble ones and the others.

The idea of existence even in heaven is more unsatisfactory to noble ones than the idea of existence in the horrifying hell is to worldlings. Then, in what world do the noble ones find contentment? The next section 5.6 describes a certain interesting aspect regarding it.

5.6 Dhyana Meditation

It is like the sun radiating toward everything in the world without restriction, unlike the firefly twinkling for its own purpose toward the companion. In fact, *dhyana* meditation is the basis of all practitioners' exercises.

The ones practicing the *dhyāna* practitioners become focused and capable of exact contemplation. The word *dhyai* is used in the sense of "contemplating exactly according to the truth." Concentration is called *dhyana* meditation only when it is excellent in the following way. Concentration has two features: *calm abiding* and *insight wisdom*.

Meditation is usually carried out in the sitting posture. Citing a passage from a sutra (*Madhyama Agama*), a bodhisattva confessed as follows (*AKB* [2, Chap. 2, §BAB 2.5]):

> *As long as I have not attained enlightenment, I will not break this cross-legged position for meditation.*

The bodhisattva does not change his posture because of the intention to attain enlightenment, for it is in a single "sitting" that he completes his goal.

During meditation, the *mindfulness of breathing* is essential. The mindfulness (*smṛti*) of breathing is a state of awareness focused on *in-breathing* and *out-breathing*, a wisdom in its intrinsic nature provoked by the power of mindfulness. In fact, it is practiced by Buddhists. The practitioners mindfully follow the air that is inhaled quietly through the nose and that goes deep into the lungs because of abdominal breathing. They also mindfully follow the exhaled air exiting quietly through the nose. Mindful breathing involves both the mind and the body working together so that the whole body can experience both the inside world and the outside world simultaneously as they are.

The bodhisattva episode given above suggests that meditation is a sequence of timewise transformation during calm abiding, not a snapshot at a single moment. *Dhyāna* meditation is a timewise transformation of the practitioner in harmony with the external world through *calm abiding* as well as the internal physical body of the practitioner, within which a physiological process equipped inherently is going on ceaselessly.

By applying mindfulness, the practitioners contemplate, "All fabricated things are impermanent; all phenomena are empty and non-self." The power of mindfulness (*smṛti*) acting in the body is like that of a wedge splitting wood. Mindfulness in calm abiding invites the wisdom of

Buddha to the practitioner. Owing to the positive power of mindfulness, Buddhist *insight wisdom* becomes active in this way together with physiological conditions such as the *runner's high.*

Wisdom develops over time; it is not a flash of insight in a single moment. One is not said to be wise at a single moment. A practitioner is said to be wise after he or she has successfully completed some practice. If the one's action was not successful, one is not said to be wise. In terms of Buddhist practices, Buddhist wisdom is concerned with whether the Noble Truths are successfully performed.

5.7 The Third Noble Truth: Cessation of *Duḥkha*

Section 5.5 described the difference between ordinary worldlings and the noble ones by using the example of an eyelash and how they look upon the mundane world — whether they view it as an eyelash placed on the palm of the hand (namely *nothing particular*) or as an eyelash getting into the eye (namely *how it leads to suffering and harm*). For this reason, the First and Second Truths are truths of the noble ones. Though the Third and Fourth Noble Truths to be presented here are respected by the noble ones, both truths are open to others freely and without restriction.

The First and Second Truths are gained by personal experiences. Likewise, the Third and Fourth Truths must be experienced and acquired by each practitioner. Specifically, those cannot be gained through theory (through brain function alone), but must be experienced by the entire body of practitioners.

How to enter the path: First and foremost, whoever desires to see the truths should guard oneself under the Buddhist precepts (according to *Abhidharmakośa-bhāṣya* [2, Chap. 6, § AD]).

> Then, they listen to the teachings on which the insight into the truths depends.
>
> Having listened, they reflect exactly. Having reflected, they devote themselves to the cultivation of *samādhi*-concentration (concentrated meditation).

Definitions of the abhidharma theory according to *AKB* (2, Chap. 4, § CAD): For the definition of the noble ones (*ārya*), see the beginning paragraph of Section 5.3. The noble ones (*ārya*) detached from the *bhavāgra* (the last stage of *saṃsāra* transmigration) are perfected beings; the noble ones detached from the realm of *desire* are non-returners (*anāgāmin*). Those who have attained the cessation of *sensation* and *conception* (of aggregates) are the ones who have acquired *nirodha* (cessation).

Non-returners who have acquired cessation are called *bodily witnesses* or those who realize within their own bodies. This is because, in view of the absence of thought, through their body alone, they have directly realized a phase similar to *nirvāṇa*, i.e. the attainment of cessation.

How do they directly realize through the body alone? Vasubandhu replies as follows (AKB[2], Ch. 6, §BAB2.4.1, p. 1960–61):

> When the noble ones leave the attainment of cessation, as soon as they think: "Oh! This attainment of cessation is peaceful like **nirvāṇa**!"
>
> They acquire a calmness of the conscious body never previously acquired (i.e. of the body in which the consciousness has arisen again). In this way, **they realize directly through the body the calmness of cessation** by virtue of two actions of realization: the first, during the attainment, the acquisition of a body in accordance with cessation; the second, upon leaving the attainment, the cognition that becomes conscious of the state of the body. For, the fact of "manifesting" is called realization. There is realization when one notices the calmness of the body that has again become conscious; and, from this noticing, the result is that this calmness has been acquired while the body was non-conscious.

Thus, we have

The Third Noble Truth: cessation of duḥkha

putting an end to craving and clinging.

In Section 2.1 of Chapter 2, it is explained that *abhidharma* denotes "direct realization of dharmas" or "putting into the true nature of dharmas." Therefore, one realizes that this section describes the central theme of the *abhidharma* theory, concerning the attainment of cessation and direct realization of the calmness of cessation.

Specifically, the *bodily witness* or *bodily verification* is the central core of Buddhists activity. In fact, *direct realization* is not only the theme of the *abhidharma* theory but also the central theme of Mahāyāna Buddhism. We will see this in various parts of the second half of this book.

5.8 The Fourth Noble Truth: The Noble Eightfold Path

Having *witnessed* the truth of cessation *realized* within one's body, let us consider the Buddha's practical instructions on putting an end to suffering, and ask what is the path to the cessation, i.e. the path (*mārga*) conducive to *nirvāṇa*. Noble ones take the path of liberation by calming down proclivities (latent defilements, Section 5.2) such as *craving, desire, clinging, hostility, dissatisfaction, afflicted view, ill-will,* or *doubt*. In fact, these die down as one follows the path.

Noble ones then take the path of insight wisdom, which is of the nature of a wheel of dharma, i.e. a wheel of the Noble Eightfold Path, which is presented, not with a rectilinear sequence, but with the following eight elements having the structure of a wheel consisting of a central core axle, radiating spokes, and an overarching rim (outer ring):

1. Right perspective (*samyak-dṛṣṭi*);
2. Right resolve (*samyak-saṃkalpa*);
3. Right speech (*samyak-vāc*);
4. Right action (*samyak-karmānta*);
5. Right livelihood (*samyak-ājīva*);
6. Right effort (*samyak-vyāyāma*);
7. Right mindfulness (*samyak-smṛti*);
8. Right concentration (*samyak-samādhi*).

These are all linked together, and each helps in the cultivation of the others. However, the eight elements of the path are grouped into three categories: (*i*) overarching wisdom (Sanskrit: *prajñā*): *right perspective* and *right resolve*; (*ii*) radiating spokes of noble ones' conduct (Sanskrit: *śīla*): *right speech, right action,* and *right livelihood*; (*iii*) mental core axle (Sanskrit: *samādhi*): *right effort, right mindfulness,* and *right concentration.* The Buddha gave his teaching "for the good of the many, for the happiness of the many, out of compassion for the world" (cf. *AKB* [2, Chap. 6, § BAC4. 2]; *Noble Eightfold Path* from Wikipedia and Tricycle [https://tricycle.org/magazine/noble-eightfold-path/]).

Right perspective is known as right understanding, or right view. "Right" means "according to facts and truths." It refers to understanding things as they are. The Four Noble Truths in fact explain things as they really are. Right perspective, therefore, is ultimately reduced to the understanding of the Four Noble Truths. This is the first step and most important step in clearing one's path of confusion and misunderstanding. Nothing can be expressed in fixed conceptual terms, and clinging to concepts provides no rigid support. This is what lies behind right perspective.

Right resolve is also known as right thought or intention. Its original Sanskrit, *saṃkalpa,* denotes "definite resolve or intention or firm determination or wish for." The practitioner resolves to dedicate himself to the pursuit of the right path, with freedom from ill-will and harm. This is called right resolve. The practitioner is determined to consider everything as impermanent, without a self, and regard everything as suffering.

Right speech, right action, and *right livelihood* are noble conducts. Practitioners seek to live in a way without harming others, avoiding lying, stealing, and committing violent acts. These noble conducts not only help achieve general social harmony, but also help the practitioners to guard themselves from being hated or criticized. They control themselves and strive for selfless actions. Right livelihood avoids causing suffering to sentient beings by cheating, harming, or killing them in any way.

Right effort generates energy and strives to maintain wholesome mental states free of delusion, to develop, increase, cultivate, and perfect them. Right effort and mindfulness calm the mind-body complex, pacifying unwholesome states, encouraging the development of wholesome states, and bringing practitioners' minds to perfection.

Right mindfulness is being aware and attentive to the activities of the body, mind, and phenomena (*dhamma*). Mindfulness is interpreted as never being absent-minded and being conscious of what one is doing. With constant awareness of phenomena as impermanent, sorrowful, and without self, mindfulness aids one to neither crave nor cling to any transitory state or thing.

Right concentration leads to mind-equanimity and mindfulness, which is mainly practiced by *dhyāna* meditation to develop clarity and insight into the nature of reality and to dispel *avidyā* (nescience), ultimately attaining *nirvāṇa*. All sensations, even of happiness and unhappiness, of joy and sorrow, disappear, leaving only awareness with a relieved feeling of being at home.

Thus, in the Noble Eightfold Path under *right perspective* and *right resolve*, decent behavior and practicing mindfulness and meditation not only put an end to craving, clinging, and dissatisfaction, but also bring practitioners to clear awareness and a sense of relief — of feeling at home.

Chapter 6

Dhyāna from Scientific Aspects

Seated Buddha (Gandhāra)#
http://www.buddha.twmail.cc/23/Daimond3-1.htm

6.1 On *Abhidharma* Philosophy

One of the essential subjects of the *abhidharma* philosophy is the practice of "*dhyāna* meditation" and concepts such as *direct realization, transformation*, and *life-force*. It is surprising to find that these do not contradict modern science but rather are consistent with its modern viewpoint. This is one of the themes of the present chapter.

However, the same *abhidharma* philosophy has a weak point, which will be discussed in the subsequent chapters. Specifically, the *abhidharma* philosophy relies on the concept of *intrinsic entity* (*dravya*) and *intrinsic nature* (*svabhāva*) of phenomena, which is in sharp contrast with that of the *śūnyatā* philosophy described later. For example, the abhidharma schools conceived eight real entities: (a) four fundamental great elements (earth, water, fire, and wind) and (b) four derivative elements (visible forms, odor, taste, and tangible forms). The abhidharma schools conceived them to be intrinsic real existences and had a concept of atom (*paramāņu*) and molecule, influenced by the Greek natural philosophy. Also, in the application of mindfulness (*smṛti*), too, they saw the intrinsic nature. The application must be of the nature of insight because insight is always required for eradicating defilements, although other schools claim that the intrinsic nature of the application of mindfulness is mindfulness itself and not insight. In this way, their views are *vagrant* and not fixed.

An analogous situation is seen in the development of modern science from Newtonian mechanics to Einstein's theory of relativity. According to Newtonian mechanics, the mass of a moving body under the gravitational force does not change, which is analogous to the intrinsic real entity of matter in the *abhidharma* philosophy. However, according to Einstein's theory of relativity, mass is a kind of form that energy takes. An elementary particle can transform into another form of energy, e.g. by self-splitting (i.e. by fission) into a combined system of the same energy, consisting of particles (of smaller mass) and photons. The term *photon* derives from the Greek word for light, *phôs*. In physics, light is an electromagnetic wave, but microscopically it is a type of elementary particle that takes discrete values of energy.

The seizing of the real entity or intrinsic nature in phenomena (dharmas) has been refuted in *Madhyamakakārikā*, the great work of Master Nagarjuna, and questioned by the Mahāyāna schools based on the *śūnyatā* philosophy. This is the subject of the next Part II. In the present chapter, we focus on the concepts of *direct realization, transformation, life-force*, and *dhyāna* from the point of view of modern science.

Observing or contemplating the internal state of the body is the application of insight and mindfulness, because it is through insight that a practitioner becomes a contemplator. According to Master Vasubandhu, author of the *Abhidharmakośa-bhāṣya* (*AKB*) (the treasury of the abhidharma and its commentary), the application of mindfulness is insightful wisdom. Mindfulness has a preponderant power. With the power of mindfulness, insightful wisdom can bring its ability into full play, just as a wedge splits wood.

The masterpiece *AKB* was translated into Chinese by Paramārtha 真諦 (499–569 CE) and by Xuan-Zang 玄奘 (602–664 CE). According to Xuan-Zang's disciple Pu-Guang (普光), the *Abhidharmakośa-bhāṣya* was hailed as the "Book of Intelligence" in India. In China, Japan, and East Asian countries, the *Kośa* had been treasured as a book of fundamental importance for Buddhist studies. The Sanskrit text had been considered lost for a long time, but it was discovered by R. Samkrtyayana in 1935 in the Tibetan monastery of Ngor and published by P. Pradhan in 1967.

We investigate meditation (*dhyāna*) in this chapter. We saw the consciousness (*vijñāna*) in Section 2.5, which are of fundamental importance in abhidharma. Vasubandhu's work is regarded as a culminating achievement in streamlining the overall structure of the exposition of the abhidharma edifice. We saw in Chapter 5 that the concept of abhidharma is centered around the Four Noble Truths. For Buddhists, the practices and aim is to reach the stage that leads to liberation (*mokṣa*) from suffering and to deep insight (*vipaśyanā*). The path leads to liberation by the meditative practice aiming at *nirvāṇa*.

6.2 What is Dhyāna

(1) *Calm abiding, insight and mindfulness*

Dhyāna meditation is the concentration (*samādhi*) endowed with certain excellent qualities [2, Chap. VIII, P. 2374]. Just as a chariot is pulled by two horses harnessed to a yoke, the meditation is pulled by calm abiding (*śamatha*) and mindful insight (*vipaśyanā*). Calm abiding and insight are not pains, but they are called "happiness", "comfort" (by a sutra). It is the way in which one practices meditation comfort, which is the excellent concentration called *dhyāna* meditation.

The practitioner usually assumes the lotus posture (*padmāsana*), a cross-legged sitting meditation pose, which was a custom from ancient India: In this pose, the back is straight, the legs are crossed with each foot placed on the opposite thigh, and the hands rest relaxed on them with the thumbs touching softly at the level of the navel. In Buddhist exercise, it is

Dhyana meditation, Gandhara Buddhist work, 1ˢᵗ–3ʳᵈ CE (Pakistan).

well known historically that the *lotus posture* is effective in calming the mind and the practitioner is led to deep meditation.

To realize concentration, there are two entryways to the cultivation of meditation on *calm abiding* (*śamatha*, tranquility). (i) Those who always delight in solitude enter through meditation on dwelling in quiet. (ii) Those who are discursive are encouraged to enter through mindfulness of breathing.

Mindfulness of breathing cuts off discursiveness because it is not directed toward the outside, for it concerns breathing. Mindfulness of breathing is carried out by counting, i.e. counting from one to ten through mere mindfulness, and then by following the flow of the air that enters and leaves one's body. Practitioners are guided to follow the inhaled air entering (sequentially) the throat, the lung, the navel, the thigh, and the shank; they follow the exhaled air exiting, traveling the length to some distance. Mindfulness of breathing is a consciousness concerned with in-breathing and out-breathing. It is called mindfulness (*smṛti*, awareness) because it is provoked by the power of awareness. Thus, the practitioner sails into the abodes of mindfulness. At this stage, the meditator changes his object of meditative focus from breathing to the advanced levels of deep concentration.

Thus, realizing calm abiding, practitioners will cultivate insightful meditation (*vipaśyanā*, meditation with insight into the true nature of reality) and *become aware mindfully*. Applying mindfulness to the *body*, *sensations*, *ideation*, and *mind*, they observe their five aggregates as *impermanent, unsatisfactory, empty*, and *non-self.*

(2) *Direct realization and uninterrupted stream of transformation*

With repeated practice of the application of insightful mindfulness in meditation, practitioners eventually gain a deep insight into the dharmas (such as the Four Noble Truths) together with physiological transformation within the body, resulting in tangible attainments. This is supported by the practitioners' burning desire for the path fueled by burning the defilements. Eventually, mindful awareness arises spontaneously within practitioners, eradicating defilements and realizing *nirvāṇa*. This is the noble stream of states within the practitioner. It is like the growth of a plant: From the seed arises the sprout, from the sprout arises the stem, from the stem arise the leaves, and so on.

Thus, practitioners sail into the path of insightful awareness, known as "direct realization" (*abhisamaya*), which is the very nature of abhidharma. This is the essence of Buddhist practice. No matter how much a practitioner has struggled and succeeded in intellectual understanding and abandoning defilements, he or she cannot attain the spiritual and physiological transformation into the noble state without going through the process of direct realization (*abhisamaya*), in which insightful awareness is acquired by "transformation" (*pariṇāma* in Sanskrit). It is essential to realize the effect of the *life stream* occurring within the practitioner.

Transformation (*pariṇāma*) is the change between prior and subsequent moments, which is the life stream (*saṃtati-pariṇāma-viśeṣa* in Sanskrit) just like the fruit arising from the seed. By the stream (*saṃtati*), we mean a sequence of cause and effect that constitutes an uninterrupted life stream. Let us consider how the fruit arises from the seed. The fruit does not arise directly from the seed itself. Indeed, it arises from the culminating moment in the evolution of the stream, which has its origin in the seed. The seed successively produces the sprout, the stem, the leaf, and finally the flower, from which the fruit is born. Consider an effect of *action*: the insightful, mindful meditation. The future effect does not arise directly from the initial action. The effect arises as a result of an uninterrupted sequence of transformations in the stream, which has its origin in the initial *action*, just like the fruit arising from the seed.

(3) *Dhyāna and insightful wisdom realize nirvāṇa*
Thus, true Buddhist meditation is accompanied by full awareness and insightful wisdom. *Dhyāna* means "to understand truly." There is necessarily an element of *insightful wisdom* in a concentrated mind, which is referred to as *prajñā*.

Verse 372 of *Dhammapada* states this aspect as follows:

> *There is no dhyāna (meditation) without prajñā (wisdom), and no prajñā is there for one without dhyāna. Only the one who has both dhyana and prajñā* (with equal weight) *can realize Nirvāṇa.* (Summary and Discussion by Dhammajoti [2, p. 42])

Dhyāna, which fulfills both calm abiding (*śamatha*) and the insight (*vipaśyanā*) of proper seeing, is called an abode of happiness in the present life, and practitioners meditate at ease.

(4) *Four meditations* (four *dhyānas*)
The practice of concentration (*samādhi*) is divided into four categories of meditation.

(i) The first meditation is characterized as wholesome concentration (samādhi) endowed with initial inquiry, characterized with investigation, joy, and pleasure. Initial inquiry necessarily accompanies investigation along with joy and pleasure, just as fire accompanies smoke along with heat and light.

The second and the third meditation are characterized through successive abandonment or replacement among the members under concentration (*samādhi*).

(ii) The second meditation is endowed with joy, pleasure, and satisfaction, with investigation replaced by satisfaction.
In the first two kinds of meditation, pleasure means the pleasure of abandonment/detachment and pleasure caused by pliancy (aptitude, flexibility, adaptability), whereas the mental state of practitioners is characterized by satisfaction in the sense that they are separated from worldly affairs and feel a (restricted) sense of liberation. When the agitation of initial inquiry and investigation has come to an end, just as a river flows smoothly after the waves have subsided, the stream of body-mind flows calmly and clearly. This is called *personal serenity.*

(iii) Being detached from joy, the third meditation is endowed with pleasure and peace, with satisfaction replaced by peace (and calm).
In the third meditation, practitioners are not agitated by joyous exaltation. They abide in the calm state characterized as *serenity* and feel pleasant and peaceful. Thereafter emerges a ripened effect produced by the successive effects of meditation.

(iv) The fourth meditation is the abode of practitioners with non-conception and cessation. The former non-conception causes the cessation of thoughts. Thus, there emerge two attainments: non-ideation (asaṃjñā) and cessation (nirodha).
The fourth meditation is *unagitated*, like a candle that is not agitated without wind in a calm space. The state of non-ideation is the absence of conception. Conception means sensing an external object and discerning its color and shape. It is through *conception* that a *name* is given to an external visual impression. In other words, conception means discerning individual fine distinctions, i.e. perceiving that something is blue, round, etc.

Ordinary worldlings mistake the state of *non-conception* for true liberation. This is not noble, because they believe that *non-conception* implies *escape* from conception. Practicing with such a desire (i.e. a defilement) for liberation is not pure. On the other hand, noble ones know that defiled practice cannot lead to true liberation and eventual enlightenment. They see attainment of non-conception as a calamity and do not desire to practice it.

The attainment of the cessation of *conception* is called *liberation*, because it averts the conceptualization of external objects, or because it averts all conditioned phenomena. Noble ones practice this attainment of cessation because they consider it as peaceful abode or concentration. The thought of practitioners is *subtle* (or *delicate*) even in the first three types of meditation in comparison to any thought associated with ordinary worldlings. In the attainment of cessation, the thought is even more subtle, called *subtle-subtle*, and consciousness is free from *conception*, while *dhyāna* breathing of the practitioner lasts uninterruptedly. In the attainment of cessation, the *subtle-subtle* conception is not the state destroyed, but that transcended: that is, staying just mindful without any specific conception, or abiding in consciousness with mindful insight.

6.3 Dhyāna Meditation: Scientific Aspects

(1) *Dhyāna sitting, pariṇāma, and direct realization*

Dhyāna meditation has many excellent benefits for the body as well as for the mind. Body health conditions such as cholesterol levels or blood pressure can be cured if meditation is performed properly and regularly.

There is a Sanskrit term, *pariṇāma*, which denotes *transformation* not only of the mind but also of the body. This section investigates how the concept of *pariṇāma* applies to the Buddhist practice of *dhyāna*. It can be said that all the phenomena (*dharmas*) occurring in our body are a sequence of the transformation of states arising inside the body deeply, which cannot be seen from outside the body. Specifically, they are physiological and autonomous functions working within our body. It is interesting to compare them with dynamical systems in physics, and they remind us of the *time-wise* evolutionary processes of mechanical systems or objects in cosmic space.

Let us consider such a sequence of transformations with an example from the natural world. Consider a piece of dried wood that is not burning.

It can yield *fire* when rubbed. In scientific terms, the rubbing action (regarded as an external stimulation or excitation) is understood to be a mechanism by which the kinetic energy of the motion associated with the rubbing action is transformed into heat energy. The heat is transmitted to the wood. Acquiring the heat, the wood reaches its ignition temperature. If that is realized, the wood begins to burn owing to the chemical reaction with the oxygen present in the air. These processes are regarded as a sequence of transformations (e.g. the temperature increase) and chemical reaction. Thus, the wood burns and gives rise to flames.

Similarly, consider a person practicing *dhyāna* sitting in the *lotus posture*. Within the practitioner's physical body, a physiological process of progressive transformations is under way. As will be explained in the following text, the *dhyāna* (or *yoga*) posture imposes certain tensional stresses on various muscles of the practitioner's physical body. This is analogous to external stimulation, such as the rubbing action that causes wood to burn. In other words, the dhyāna (yoga) meditation under tensional stress is analogous to being "restrained to a yoke (or harness)." The yoga posture imposes a certain stimulative stress on the physiological condition of the practitioner. (Note that the term *yoga* is derived from either of two roots, *yujir yoga* [to yoke] or *yuj samādhau* [to concentrate].) The ultimate goal of *yoga* is *mokṣa* (internal liberation) by means of the technical practice of controlling the body and the mind.

Responding to the external stimulus, the practitioner's body under internal stress responds autonomously and tries physiologically to cure the pain (even if it is faint) caused by the stress. There are physiological mechanisms in the living body under stress to release bioactive substances: *hormones* such as adrenaline and cortisol, or *neurotransmitters* such as *endorphins, serotonin, and dopamine*. These cause internal transformation within the practitioner's body. In fact, a hormone is one of signaling molecules (produced by endocrine glands) and is transported by the circulatory blood system to the distant target organ to regulate the physiological condition. As a result, the practitioner's physiological condition is modified after a while as the practice proceeds, tending to a new state of reality of the body-mind as a whole. If the conditions are favorable, this helps the practitioner to attain a state of cessation. This interpretation is also applied to the yoga practice. It is believed that all the processes are a sequence of transformations that progress within the practitioner's body. These were unknown to the practitioners in old times but amazingly were experienced as direct realization (*abhisamaya*) and captured as an uninterrupted stream of transformations (*pariṇāma*).

(2) *Lotus posture (padmāsana) and abdominal breathing*

In the lotus posture (padmāsana), the back is straight, both legs are crossed, and the feet are placed on the opposing thighs. In this posture, the practitioner's mind is calmed gradually, and the practitioner is led to deep meditation in about 10 minutes. This posture is mechanically stable from the *physics* point of view, because the body is supported by three points triangularly, with the knees touching the floor mat and the bottom (buttocks) placed firmly on a cushion. In this way, the practitioner remains stable. In this posture, breathing slows down, muscular tension decreases, and blood pressure subsides. Thus, the practitioner can maintain the posture steadily for 30 to 60 minutes, and it is thought to be the best sitting posture for meditation.

Lotus posture, Gandhara buddhist work, 1st–3rd CE (Pakistan).

In the lotus posture, the practitioner feels stress at both knees, ankles, and hips, and the back is intentionally maintained straight. Although maintaining this posture imposes some stress owing to the muscle strains at the knees, ankles, hips, and spine, practitioners may not feel unpleasant; in fact, they feel very comfortable when they get accustomed to this position. Let us consider how this happens.

In padmāsana, the practitioner practices with deep breathing (abdominal breathing or diaphragmatic breathing). This is carried out by pushing down the abdomen. Deep breathing is marked by the expansion of the abdomen rather than the chest when inhaling. Abdominal breathing helps to lower blood pressure, and the resulting release of *serotonin* makes one feel good.

This abdominal breathing (Fig. 6.1) moves the abdomen down and up as one inhales and exhales, pushing the abdominal organs forward and then back inward. The air is directed into the lower lobes of the lungs and the organs in the lower torso are stimulated, massaged, and relaxed. This improves mental focus and the practitioner acquires clarity in mind owing to the increased blood flow to the pre-frontal cortex in the brain. In addition, sleep quality is improved.

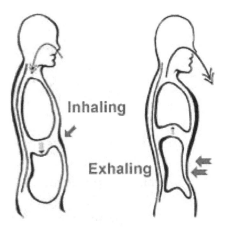

Fig. 6.1 Abdominal breathing. From "Abdominal Breathing for Heart Disease" (Yoga-adviser, 2016) [17].

In addition, the endocrine glands in the internal organs (viscerals) are activated by the massaging movement and produce a variety of hormones enhancing a wide range of physiological processes, including *stress response, immune response,* and regulations of *metabolism, blood pressure,* and *salt concentration.* Our health and emotional state are controlled or influenced significantly by the performance of those functions. The abdominal *up-and-down* motion can lead to better performance of those organs.

Abdominal breathing for 20 to 30 minutes reduces anxiety and stress. Deep breathing increases the supply of oxygen to the brain and stimulates the *parasympathetic* nervous system promoting a state of calmness. Breathing helps one feel connected to one's body and liberates one from worries, calming the mind. As the yogi (proficient in lotus sitting and abdominal breathing) follows the proper rhythmic patterns of slow deep breathing, the movement strengthens the respiratory system, soothes the nervous system, and reduces craving. As desires and cravings subside, the mind becomes unconstrained and eventually liberated.

According to these observations, Buddhist philosophy is not essential in this exercise; the yoga practice has nothing that is specifically Buddhism specific. In other words, one who practices yoga for the cessation of *dukkha* (suffering) may be called a Buddhist. Thus, it is said that Buddhist practices are scientific but that Buddhists are not necessarily scientists.

(3) *Physiology and stress mechanism under the sitting, leading to runner's high*

While maintaining the *dhyāna* (or *yoga*) pose, the practitioner is subjected to some stress and feels dull pain (sometime feeling pleasant to one's surprise), because tensional stresses are imposed on various parts of the muscle of the practitioner's body. However, the imposed stress does not necessarily lead to pain; rather in certain normal conditions, the practitioner feels not only decreased levels of pain but sometimes comfort, like in the *runner's high* explained in the following text.

Stress and pain are two most common factors that lead to the release of *endorphins* (one of the neurotransmitters in the brain). *Endorphins* interact with the *opiate*-receptors in the brain to reduce the perception of pain, acting like drugs such as *morphine*. In contrast to the opiate drugs, activation of the *opiate*-receptors by the *body's endorphins* does not lead to addiction or dependence. (The endogenous opiate system is body's internal system for regulating pain, reward and addictive behaviors, and consists of opiate substances produced naturally within the body and their receptors, into which opiates fit like keys into locks.)

In addition to the decreased feeling of pain, the secretion of endorphins leads to the feeling of *euphoria* and enhancement of the *immune* response. Hence, with high endorphin levels, we feel less pain and fewer negative effects of stress. Endorphins are considered as a modulator (an agent transforming the physiological state) of the so-called "*runner's high*" that athletes achieve with prolonged exercise. It is known that the body produces endorphins in response to prolonged, continuous exercise. The sitting meditation is considered to be analogous to such exercises.

Certain foods, such as chocolate and chili peppers, can also lead to enhanced secretion of endorphins. In the case of chili peppers, the spicier the pepper, the more endorphins are secreted. Release of endorphins upon chocolate consumption likely explains the comforting feeling that many people associate with it and the *craving* for chocolate when stressed. In order to increase the body's endorphin levels, strenuous exercises are not always necessary. One can also try various activities, such as *massage therapy* and *yoga*, which may stimulate endorphin secretion.

Thus, it is quite possible that the practice of meditation can enhance endorphin secretion and that the practitioner experiences less pain and fewer adverse effects of stress and thus feels *pleasant*. Certain essential elements of yoga are important for both Buddhism in general and Zen in particular.

(4) *Deep breathing and prolonged exercise: Physiological facts*
Deep breathing during meditation is a *stress reliever* that results in deep rest and relaxed responses to stress physically and emotionally, such as lowered *heart rate*, *blood pressure*, *breathing rate*, and *muscle tension*.

In regard to the dhyāna meditation, deep breathing leads to some desirable effects resulting from the physiological processes occurring within the practitioner's body, which are stimulated by body exercise. Three mechanisms are worth mentioning.

The first is the connection between the gut and the brain, which participate during the exercise. The internal organs in the abdomen are called the second brain, because they control the physiological condition of the whole body in collaboration with the upper main brain part.

The second is the limbic system, which is a vital part of the immune system and also protects the body by repelling bacteria and viruses. Deep breathing has the potential to activate an important limbic organ located in the abdominal region. Hence, it contributes to enhanced immunity and body cleaning.

The third is the secretion of serotonin, a neurotransmitter that contributes to pleasant feeling and helps overcome depression.

6.4 *Deep Breathing* (Abdominal Breathing): Physiological Aspects

In this section, we discuss the remarkable benefits of deep breathing for our body and mind from two points: (i) Abdominal deep breathing activates the structure of the human nervous system of the *gut–brain* connection and the function of the second brain (the abdominal part that consists of internal organs); (ii) Abdominal deep breathing helps our lymphatic system protect our body (the immune function) from infections and other harmful invaders such as bacteria and viruses and remove waste from every cell while helping the immune system.

(i) *Gut–brain connection and the second brain*
During deep breathing (or abdominal breathing), the abdomen moves up and down with each inhalation and exhalation, pushing the abdominal organs outward and then back inward, pushing the abdominal organs forward and then back inward. This exercise activates a number of internal organs located in the abdomen, such as the *stomach*, large and small

intestines, liver, adrenal gland, pancreas, kidney, and *spleen,* the important organs that control the physiological conditions of the entire body. A recent physiological study has identified the abdominal region containing these organs as an important system connected with the brain and calls it the *second brain.* It is highly likely that the *up-and-down* motion of the abdomen activates these organs and improves their performance. This scientific point of view is of *utmost* importance in regard to the *dhyāna* meditation. Let us show the evidence.

Hidden within the walls of the abdominal organs is a "brain" in the gut that links the gut, various internal organs (that control mood and health), and the main brain (Fig. 6.2).

Here is an example that shows their important functions: the *adrenal glands* (located above the kidneys) are endocrine glands that produce a variety of hormones, including *adrenaline, steroids,* and *cortisol. Corticosteroids* are involved in a wide range of physiologic processes, including *stress response, immune response,* and regulation of *inflammation, metabolism, blood pressure,* and *salt concentration.* Our *health* and *emotional* state are controlled or influenced significantly by these functions. The *up-and-down* motion of the abdomen can lead to better performance of the aforementioned internal organs. It might not be an exaggeration that Buddhist practices have verified this in the past for more than 2000 years.

In the second brain, there are nerve cells called the *enteric nervous system* (ENS, Fig. 6.2). The ENS is two thin layers of more than 100 million nerve cells that line our *gastrointestinal* tract (from the *esophagus,* the connection between the throat to the stomach, to the *rectum,* the final section of the intestine). The human nervous system can be activated in response to environmental factors such as emotion and stress. Human body responds to stresses by activating a wide array of behavioral and physiological responses, collectively referred to as the *stress response.* The CRF-hormone (corticotropin-releasing-factor) plays a central role in the stress response by regulating the hypothalamic-pituitary-adrenal (HPA) axis (Fig. 6.2).

In particular, the HPA axis (dashed line) is activated by emotions or stress. HPA activation results in the release of *cortisol* from the adrenal glands, driven by a complex interaction in the brain.

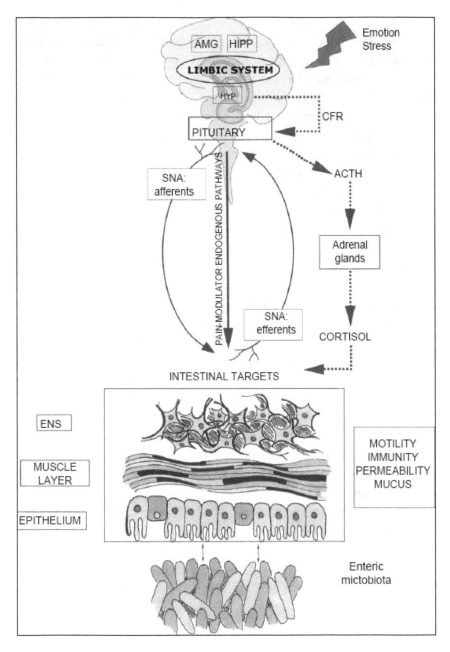

Fig. 6.2   Gut-brain connection [19].

Simultaneously, the central nervous system (CNS) communicates with the intestine organs (IO) along both the SNA-afferent (from the CNS to the IO) and the SNA-efferent (from the IO to the CNS) pathways. Thus, the CNS communicates with different intestinal targets such as the ENS *a*, muscle layers *b*, and gut mucosa, modulating motility, immunity, and secretion of mucus. The enteric microbiota *c* has a bidirectional communication with these intestinal targets, modulating the gastrointestinal functions and being itself modulated by the brain-gut interactions.

The HPA axis participates in the integration of adaptive responses to stress. The neuronal and endocrine systems contribute to the regulation of the HPA axis and the *maintenance of homeostasis* in the face of aversive stimuli.

(ii) *Immunity of the lymphatic system, and body cleaning by deep breathing*

The immune function depends on the condition of the lymphatic system. The lymph nodes is the checkpoints of the immune system. They are major sites of lymphocytes and white blood cells that protect the body by repelling foreign cells, bacteria, and viruses. Hence, they are important for the proper functioning of the immune system and act as filters for foreign particles. The abdomen contains a large number of lymph nodes, which are activated by deep breathing.

Furthermore, the upper part of the abdomen contains *receptaculum chyli* (RC in Fig. 6.3, or *cisterna chyli*), a dilated sac found near the lower area of the thoracic duct in the lumbar region of the abdominal cavity where lymphatic fluids (fluids from the body tissues associated with living activities) converge from the lower parts of the body on the way to the subclavian vein. Receptaculum chyli collects and temporarily holds the lymphatic fluids. It is shaken by deep breathing and thus activates the circulation of the lymphatic fluids. Certainly, abdominal breathing activates the function of the lymphatic system.

Figure 6.3 shows the network of the lymphatic system in our body. The *lymphatic fluid* circulates in the lymphatic system. It initially leaks out of the arteries because the blood in the arteries is under a higher

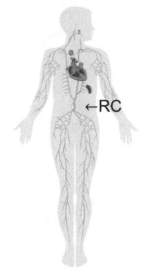

Fig. 6.3   Lymphatic system and *receptaculum chyli* (RC) [18].

hydrostatic pressure than that of outside. As it gets collected in the tissues together with the body wastes, both are sucked up by the lymphatic system and returned to blood circulation at the subclavian vein. The lymphatic system helps remove toxins and pathogens from the blood and interstitial fluid and returns the fluid and solutes from the interstitial fluid to the circulatory system. Thus, the lymphatic system cleans the body. Activation of *receptaculum chyli* by breathing is good for the cleaning and protection of the body.

The lymphatic system is a part of the circulatory system and a vital part of the *immune* system, comprising a network of lymphatic vessels that carry a clear fluid called lymph (from Latin, *lympha*, meaning "water") directionally toward the heart. Unlike the cardiovascular system, the lymphatic system has no driving pump. Consider the muscular motions of stretching (thinning) and contraction (thickening) by body exercise. Suppose a lymphatic vessel containing lymph is located between the muscle fibers. Then, lymph in the lymphatic vessel is driven indirectly by the muscular motion due to body exercise. Thus, muscular motions are important for the circulation of lymph. It is observed that deep breathing is good for lymph circulation.

The lymphatic system is not a closed system. The human circulatory system processes an average of 20 liters of blood per day through capillary filtration, removing blood plasma (including hormones), out of which three liters remains in the interstitial fluid. One of the important functions of the lymphatic system is to provide a return route to the remaining three liters of blood and simultaneously to clean the interstitial fluid between cells. Another important function is that of defense in the *immune* system. Lymph is very similar to the blood plasma, which contains lymphocytes and other white blood cells. It also contains waste products and cellular debris together with bacteria and proteins.

6.5 Repetitive Prolonged Exercise

Here we discuss the remarkable benefits of the repetitive prolonged exercise for our body and mind from three points: (i) serotonin secretion and pleasant feeling, (ii) walking meditation, and (iii) practical aspects of deep breathing and walking meditation.

(1) *Serotonin and pleasant feeling*
A recent neuroscience study suggests that prolonged rhythmic exercise increases brain serotonin function. Sitting meditation (with periodic deep breathing) and walking meditation could be included in such prolonged exercise. Immersed in their exercise, practitioners feel pleasant by overcoming the sense of exhaustion.

Serotonin, one of the neurotransmitters (which carry impulse signals between nerve cells), is thought to regulate anxiety, happiness, and mood in the brain, while it controls the movement of the stomach and the intestines in the abdominal region [20]. Abdominal breathing helps in *serotonin* release. Serotonin is thought to be involved in constricting muscles smoothly, regulating repetitive body exercises and contributing to well-being. When serotonin levels are normal, we feel *happier, calmer, more focused, less anxious, and more emotionally stable*. The following factors boost serotonin levels within the body: (a) meditation, (b) prolonged repetitive body exercise, and (c) exposure to bright light.

Our brains have evolved over the past 500 million years. Located deep inside the brain, there is an organ called the *brain stem*, sometimes called the *primitive brain* because it resembles that of reptiles. This part handles basic life support — *breathing, heartbeat*, and *digestion*. It keeps us alert and ready to react to the incoming sensory information.

The brain stem contains the *raphe nuclei* (*seam* in Greek), a tiny cluster of nuclei that release serotonin to the rest of the brain. The raphe nuclei provide feedback to the *suprachiasmatic* nucleus (SCN, Fig. 6.4), which regulate the *circadian* rhythms (roughly 24-hour cycle) in animals. The SCN transmits to the raphe nuclei the signal to alter serotonin levels for the sleep/wake states. The raphe nuclei then transmits the feedback to the SCN about the animal's levels of alertness.

Although serotonin is well known as a brain neurotransmitter, it is estimated that 90 percent of the body's serotonin is made in the digestive tract. In fact, certain bacteria in the gut play an important role in the production of peripheral serotonin, and altered levels of this peripheral serotonin are linked to diseases such as irritable bowel syndrome, cardiovascular disease, and osteoporosis.

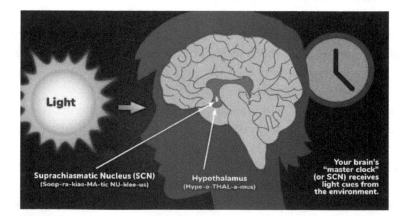

Fig. 6.4 The master clock coordinates biological clocks from the received light. Credit: NIGMS, NIH, USA.

Neurobiologists have proposed the following interpretation from the viewpoint of evolutionary human history [21]:

"There has been a large change in the level of vigorous physical exercise experienced since humans were hunter-gatherers or engaged primarily in agriculture. In later generations, decline in vigorous physical exercise and, in particular, in effort-based rewards may contribute to the high level of depression in today's society. The effect of exercise on serotonin suggests that **the exercise itself may be more important,** *rather than the rewards that stem from exercise. If trials of exercise to prevent depression are successful, then prevention of depression can be added to the numerous other benefits of exercise."*

This suggests that serotonin may have anti-depressive effects. In the next subsection, let us consider the walking meditation from this point of view.

(2) *Walking meditation*

Walking meditation is more than just a leisurely walk. It is usually performed at a much slower pace than the normal walk and involves either mindful coordination with breathing or specific focusing practices. It looks more like meditation than like walking. The art of walking meditation is to be aware as we walk. Using the walking movement, the practitioner cultivates mindfulness and wakeful presence. It can be practiced regularly before or after sitting meditation, anytime and anywhere [22].

The practitioner walks slower than usual with a relaxed pace avoiding discursive attention, opening one's senses to the surroundings and feeling stretching at in the legs and pressure on the soles of the feet. Paying attention to every part of body with each step, the practitioner feels the sensations of lifting one's foot and leg off the earth and then mindfully places one's foot back down, feeling each step mindfully as walking.

The Vietnamese Zen Buddhist Ven. Thich Nhat Hanh has adopted a simplified approach to walking meditation and imparted fresh momentum to the mindfulness movement in the modern world. According to his interpretation [23]:

> When we practice walking meditation, we arrive at our home in each moment. Our true home is in the present moment. When we enter the present moment deeply, our regrets and sorrows disappear, and we discover life with all its wonders. Breathing in, we say to ourselves, "I have arrived". Breathing out, we say, "I am home". When we do this, we overcome dispersion and dwell peacefully in the present moment, which is the only moment for us to be alive.

Walking is regarded as prolonged rhythmic exercise, discussed in the previous subsection, *Serotonin* and *pleasant feeling*. Hence, one can expect serotonin secretion in the brain during the walking exercise.

Dialogues between Sri Lankan Buddhist philosopher Ven. Alubomulle Sumanasara Nayaka Thero and Japanese brain scientist Prof. Arita have been published in the Japanese book *Buddhism and Brain Science* ([24], 2010). According to Ven. Sumanasara, walking meditation is one of the practices in his Theravada school. During walking meditation, each practitioner does "live-reporting" of his/her own walking practice silently, in order to focus on the present moment and be aware to the on-site exercise. This is simply to report the movement of alternating steps with the left and right foot mindfully. After live-reporting for a while, the practitioners acquire full awareness and are transformed into a *samādhi* state spontaneously. The walk of live-reporting is different from the walk involving discursive attention. See the next subsection.

(3) *Practical aspects of deep breathing and walking meditation*

It is known that everyday exercise of abdominal breathing (deep breathing) for 10 to 30 minutes reduces anxiety and stress. Deep breathing boosts oxygen supply to the brain and stimulates the *parasympathetic* nervous system, which promotes a state of calmness.

Breathing practices help one feel connected to one's body, bring awareness, remove worries, and calm the mind. By practicing deep breathing, the heart rate slows down, the muscles relax, breathing becomes slower, and the blood pressure decreases [30]. Following is an American practitioner's experience [25]:

"In 2009 I had the privilege to be able to attend a retreat in the mountains of Colorado with the ordained sangha of *Thay* (respectful reference to Ven. Thich Nhat Hanh). Early every morning we gathered before the sun rose and practiced walking meditation together. Over two-hundred people, in slow motion, made their way to a clearing field, sat down there, enjoyed the crisp air, and witnessed the sunrise. Then we slowly stood up, stretched our legs, then mindfully walked to the main hall for morning meditation and teachings. I aspire to bring to daily life what I learned from it. As I walk to check the mail, as I open a book to read, as I sit and type at the computer, I strive to bring *mindfulness* to every moment, to each breath. It's good to slow down, to realize that each step we take can be a step toward freedom for all sentient beings. Peace is possible in every step we take. Peace is possible in every moment."

6.6 Gut Microbiota (or gut flora) — *Note added in proof*

Our human beings (like most animals) have a certain external *microbiota* within our bodies, which are not completely controlled by our nervous systems. Those are microorganisms including bacteria, viruses and others that live in the digestive tracts. The relationship with some of those gut microbiota is not merely non-harmful coexistence, but rather a mutualistic relationship. But there are some others which are harmful and (say) cause diarrhea.

It is vitally important for lively daily life of ourselves to keep friendly and healthy **coexistence** with those huge amount of tiny others living within our bodies, estimated to contribute more than 1 kg of our bodyweight.

Verse praising Zazen Sitting Meditation
by Fourth-Zu Daoxin (580–651)

Observing one's mind during Zazen Sitting, its nature is freely-streaming, never stagnant.

Cleanly empty and peacefully serene, body and mind settling down properly.

Feeling at ease, gently inhaling, deeply exhaling.

Breathing air, cool and clean mind settling down gradually.

Spiritual way, clear and sharp, mind state, bright and clean.

Observing it, obvious and clear, Inside and outside, empty and clean

Mind nature, Nirvana.

As is the nirvana, holy spirit manifests itself.

Though it is of no form, minding exists constantly, and spiritual soul does not run out.

Always existing, bright and clear, it is named Buddha nature.

Those seeing the Buddha nature liberated eternally from worldly affairs.

四祖法話 (四祖 大医道信 大師, 580–651)

坐禅看心　　滔然得性
清虚情静　　身心調適
能安心神　　窃窃菓冥
気息清冷　　徐徐叙心
神道清利　　心地明浄
観察分明　　内外空浄
心性寂滅

如其寂滅　　聖心顕実
性雖無形　　志節煙在
　　　　　　幽霊不渇

常存朗然　　是名佛性
見佛性者　　永離生死

(Fourth-Zu Daoxin: See §10.5.4)

Chapter 7

Last Strong Message of Shaykyamuni:
Tathagata

Ancient Buddhist Concept and Modern Biology

More than 1500 years ago, Buddhists had an insight in the substratum *ālaya*, eighth consciousness, also called *Tathāgata-garbha*, considered as potential seed for mental and physical storehouse of one's existence, on which other seven consciousnesses are evolving. This is something conceptually analogous to the function of the genome of modern biology, because the eighth consciousness is characterized as a substratum, seed, or storehouse. Let us investigate the similarity and dissimilarity of both of them. Of course, the genome of modern biology has nothing to do with the *ālaya* consciousness.

Another comparison between ancient Buddhist concepts and modern science, worth detailed investigation, is the pair "Tathāgata" and "synaptic plasticity" in the brain during sleep. This is presented in detail in Sections 3.6–3.8.

7.1 *Tathagata-garbha* and *Genome*

There are eight consciousnesses in the Buddhist philosophy according to Section 2.4 (vi): five sense consciousnesses (associated with the five sense organs); the sixth mind (mental consciousness); the seventh *manas* (self-consciousness); and eighth, substratum, or *ālaya* consciousness. The eighth consciousness forms the base consciousness. The other seven consciousnesses are "evolving" or "transforming" consciousnesses on the base consciousness. The eighth consciousness is called storehouse consciousness because it is characterized by accumulation of all potentials as seeds for one's mental and physical existence. Since it serves as a container for all experiential impressions, it is also called the "seed consciousness." It is in fact the potential consciousness that induces origination of a new existence.

The idea that all sentient beings are seeded with the Buddha-nature (and thus are potential Buddhas) is accepted widely among Buddhists. Terms used to designate this concept are *ālaya* consciousnesses, Buddha-nature, *Tathāgata-garbha*, *Tathāgata essence*, or *Tathāgata storehouse*. These are the recognition of Buddhists, originating from the Shakyamuni (Sage of the Shakya clan), Gautama Siddhārtha, himself. In fact, the Shakyamuni is said to have called himself Tathāgata. This recognition is the outcome gained historically through long-term practices and experiences of all Buddhists.

The Sanskrit *garbha* denotes " germ of a seed," "womb," "embryo," "conception," and "containing inside" (i.e. something within which something else originates or develops). The word *garbha* also means "essence" or the "essential nature," which gives us the phrase "Buddha-essence." It would be worth paying attention, in particular, to the naming the eighth consciousness as *Tathāgata-garbha* rather than *Buddha-garbha*. Section 7.2 rephrases what is meant by Tathāgata and finds another profound meaning.

Needless to say, the genome of modern biology has nothing to do with the word *Tathāgata-garbha*. The term *genome* was created in the 20th century and is a blend of the words *gene* and *chromosome* in molecular biology. Within biological cells, the DNA molecule is packaged into thread-like structures called chromosomes. Each chromosome is made up of DNA tightly coiled many times around proteins supporting its structure.

The term *genome* means "all genetic information of an organism," consisting of sequences of DNA (deoxyribonucleic acid). The genome is stored in long molecules of DNA as a structure called a double helix. The DNA is made up of chain units of biological building blocks of four letters of base molecules. The DNA contains instructions that are necessary for our body to grow, develop, and reproduce. These instructions are stored within the sequence of base pairs. The DNA functions as a storage device for information having enormous potential.

The last description (given above) of the DNA (or a genome), containing the instructions that are necessary for our body to grow, develop, and reproduce, is very similar to the Buddhist expression concerning the eighth consciousness, *Tathāgata-garbha*, which does function like the germ of a seed, i.e. the one within which something else originates or develops. In addition, the *Tathāgata-garbha* is called storehouse consciousness because it is characterized by accumulation of all potentials as seeds for one's mental and physical existence. It is, as it were, a potential consciousness that induces origination of a new existence. Correspondingly, the genome contains all genetic information as a storage device for future development and biological reproduction.

However, it is emphasized that the connection between Tathāgata practice and synaptic plasticity in the brain during sleep has more realistic and physiological significance, which is considered in the next sections.

7.2 *Tathāgata*

Let us investigate an insightful claim: "The essence of the message of Shakyamuni might be expressed by the *Tathāgata*. What is meant by the term *Tathāgata*?".

(a) *Etymological* consideration of the *Tathāgata*

In the Buddhist Pāli canons, the Shakyamuni himself used the term *Tathāgata* when referring to himself instead of using the pronouns *me*, *I*, or *myself* (an example is given below). The term *Tathāgata* is usually used to denote Buddha. However, from an etymological point of view, there exists some interesting difference between them: Buddha means "awakened," while Tathāgata means "one who has thus come," or 如来 in Chinese and Japanese.

The word *Tathāgata* is a Sanskrit compound consisting of the terms *tathā* (thus) and *gata* (gone), or *āgata* (come). The term *Tathāgata* is often thought to mean either "one who has thus gone" (*tathā-gata*), or "one who has thus come" (*tathā-āgata*). After the citation given below, another interpretation is given in the light of modern science.

The difference between Buddha and *Tathāgata* might be stated as follows: "awakened" denotes one's present state, while the "one who has thus come" is a timewise description. In fact, in response to the question "Why is the Buddha called *Tathāgata*?" a Mahāyāna text states: "In the way that the previous Buddhas have gone by the path of Nirvāṇa, thus (tathā) the actual Buddha is going (gata) and will not go on to the transmigration." Hence, the Tathāgata is an expression inseparable from the path of *nirvāṇa* and the Noble Path.

Note that, in the Mahāyāna Buddhism, there is the concept of Eternal Buddha transcending through time and commonly associated with Shakyamuni Buddha, but can also refer to both his past and future incarnations.

> "*Tathāgata*" is defined as someone who "knows and sees reality as-it-is". *Gata* "gone" is the past passive participle of the verbal root *gam* "go, travel". *Āgata* "come" is the past passive participle of the verb meaning "come, arrive". In this interpretation, Tathāgata means literally either "the one who has gone to suchness" or "the one who has arrived at suchness".
>
> The sage cannot be "reckoned" because he is freed from the category "name" or, more generally, concept. The absence of concept precludes the possibility of reckoning or articulating a state of affairs; "name" here refers to the concepts or apperceptions that make propositions possible.
>
> ["Tathāgata," Wikipedia]

(b) What is implied for the Tathāgata in Pāli texts: From *Maha-parinibbana Sutta*

The Pāli canon *Maha-parinibbana Sutta* [13] records Buddha's final messages, including the term *Tathāgata*. In this text, one can find expressions regarding "What is implied by the Tath*ā*gata?" This text describes the stories of his last journey from Magadha to Kushinagar with a number of important and valuable dialogues that were exchanged between Shakyamuni and his disciples, *bhikkhus* ([13] Japanese edition, Chap. 2, p. 62–64). One of them is as follows:

Ananda, the disciple bhikkhu, spoke to the Shakyamuni:

Bhagavat (Venerable, Shakyamuni), I still had some little comfort in the thought that Bhagavat would not come to his final passing away until he had given some last instructions respecting the community of bhikkhus (Sangha).

Shakyamuni answered him, saying:

What more does the community of bhikkhus expect from me, Ananda? I have set forth the Dhamma without making any distinction of "inner" and " outer"; there is nothing, Ananda, with regard to the teachings that the Tathāgata holds with the closed fist of a teacher who keeps some things back. Whosoever may think that it is he who should lead the community of bhikkhus, or that the community depends upon him, it is such a one that would have to give last instructions respecting the community. But, Ananda, the Tathāgata has no such idea as that it is he who should lead the community of bhikkhus, or that the community depends upon him. So what instructions should he have to give respecting the community of bhikkhus? Why should the Tathāgata make arrangements for the community?

where dhamma denotes "way of living" in Pāli, corresponding to *dharma* in Sanskrit. Subsequently, Shakyamuni expressed himself from his heart:

I am now frail, Ananda: old, elderly, advanced in years, having come to the last stage of life, now I am eighty years old. Just as an old cart is held together with straps, so the body of Tathāgata is kept going only with supports.

Then Shakyamuni expressed himself when he felt satisfied, as if squeezing it out:

Ananda! Only when the Tathāgata is not attending to anything at all, with the cessation of feelings, and abides in the formless concentration of mind, then his body is more wholesome.

Note that, in this one of the most esteemed Buddhist texts, the Shakyamuni called himself *Tathāgata*. The following is a well known and profound message of the *Tathāgata*.

> *Therefore, Ananda, be islands unto yourselves, be refuges unto (do rely on, [13] Japanese version) yourselves, seeking no external refuge; with the Dhamma as your island, the Dhamma as your refuge, seeking no other refuge.*

In addition, addressing to any bhikkhus devoting themselves for practices,

> *Those bhikkhus, seeking an island and a refuge in themselves (relying on themselves, [13] Japanese ed.) and in the Dhamma and nowhere else, whoever, those bhikkhus of our Sangha will surely be attaining and abiding in the highest.*

This is a dialogue during the last journey when the Shakyamuni and the bhikkhus took their rain retreat together in the village of Beluva not far from Vesali. However, in the rainy season, there arose in Buddha a severe illness, and sharp and deadly pains came upon him. Shakyamuni suppressed the illness by strength of will, resolved to maintain the life process, and lived on, thinking that it would not be fitting if he came to the final passing away without addressing the attendants, without taking leave of the Sangha community of bhikkhus. Shakyamuni recovered from the illness; soon after his recovery, he came out from his dwelling place and sat down in the shade of the building, on a seat prepared for him. The bhikkhu Ananda approached the Buddha, respectfully greeted him, and sitting down at one side, he spoke to Shakyamuni. This was the setting for what happened above.

One day, after the forenoon alms to and from Vesali, the Tathāgata, who was wearing a robe and carrying a bowl, looked upon Vesali with the elephant's look, and said to Ananda to leave for Bhandagama. Then, together with the Sangha community of bhikkhus they took up their abode there. The Tathāgata often gave counsel to the bhikkhus as follows:

> *Such and such is virtue (practice free from taint); such and such is concentration; and such and such is wisdom. When the concentration is*

fully developed by virtuous practice, the fruit gained is great and its merit is great; When the wisdom is fully developed by concentration, the fruit gained is great and its merit is great; The mind that is fully developed with wisdom is liberated completely from taints of lust and nescience (ignorance).

They took their next abode at Hatthigam, and then at Jambugama. At each of these places, the Tathāgata often gave counsel to the bhikkhus. When Tathāgata had stayed at Jambugama as long as he pleased, he spoke to Ananda: "Come, Ananda, let us go to Bhoganagara."

Thus, they continued their journey. After some days, they came close to the last place Kusinara. The Tathāgata addressed Ananda, saying: "Come, Ananda, let us cross to the farther bank of the Hiraññavati, and go to the Sala Grove of Malla-clan in the vicinity of Kusinara." Arriving there, Tathāgata spoke to the Ananda,

Please, Ananda, prepare for me a couch between the twin sala trees, with the head to the north. I am weary, Ananda, and want to lie down.

Then, the Tathāgata lay down on his right side, in the lion's posture, resting one foot upon the other, and so disposed himself, mindfully and carefully.

After a while, Tathāgata addressed the bhikkhus, saying:

*It may be, bhikkhus, that one of you may be in doubt or perplex as to the Buddha, the Dhamma, or the Sangha, **the path or the practice**. Then please question, bhikkhus! Do not be given to remorse later on with the thought: 'The Tathāgata was with us face to face, yet face to face we failed to ask him.*

When this was said, the bhikkhus were silent. Yet, for a second and a third time the Buddha repeated the same message to them. For a second and a third time, too, the bhikkhus were silent. Then, Buddha addressed the bhikkhus ([13] Japanese version Chap. 6, p. 156–8):

The Tathāgata knows for certain that, among this community of Bhikkhus, there is not even one bhikkhu who is in doubt or perplexity as

to **the Buddha, the Dhamma, or the Sangha, the path or the practice**. *For, Ananda, among these five hundred Bhikkhus even the last (immature, but eventually liberated), is a stream-enterer who is assured and bound for the highest liberation.*

Finally, the Buddha addressed the Bhikkhus, saying:

Now listen, Bhikkhus, to what the Tathāgata now says: All fabricated beings are subject to decay and impermanent. Strive for perfection through heedfulness! (Be diligent in your effort.)

These were the last words of the ***Tathāgata***.

It is particularly remarkable that the Tathāgata emphasized the **Path** and the **Practice** as well as Buddha, Dhamma, and Sangha. The latter three are usually called Triple Gems (Three Jewels) or Three Refuges, which are recited daily among Buddhists. Thus, it may be said: the temporal aspect of the term *Tathāgata* is represented by the first two: Path and Practice.

7.3   Brain Working 24 Hours: Sleep Not a Simple Lazy Rest

In this section, we consider the significance of the Buddhist Practice from the modern science viewpoint. Historically, sleep was thought to be a passive state. However, modern biological science of the 21st century is revealing a number of surprising aspects of our beings, in particular concerning our sleep, sleep is not a simple lazy rest. The ancient Buddhist philosophy *abhidharma* is mainly concerned with consciousnesses. Only the eighth *ālaya* consciousness is concerned with the substratum consciousness, called variously as the consciousness of either storehouse, base, seed, or potential consciousness. In fact, the brain is working for twenty-four hours. Scientifically, sleep is a dynamic process, and our brains are active during the time of sleeping too.

We learned in Section 3.4(b) (brain area related to *consciousness*) that the thalamus located deep inside the brain plays an important role in regulating states of sleep and wakefulness. Hence, in order to contemplate on what is sleep, why or how we sleep, one has to learn from the modern neuroscience.

(a) *"Be diligent in your effort!"* Rephrasing what the Tathāgata told bhikkhus lastly

Surprisingly enough, this has support from modern science, concerning memory consolidation during the sleep. Before seeing it in detail, let us first recall what was told by Shakyamuni on the practice. In the last section, Shakyamuni first assured the bhikkhus attending to him, saying,

> *There is nothing with regard to the teachings that the Tathāgata holds with the fist closed in which some things are kept back.*

Shortly thereafter, in the last moments, Shakyamuni addressed the bhikkhus:

> The *Tathāgata* knows for certain that, among this community of Bhikkhus, there is not even one bhikkhu who is in doubt or perplexity as to the Buddha, the Dhamma, or the Sangha, *the path or the practice*. Those bhikkhus of the Sangha will surely be bound for enlightenment."

And

> *All dhammas of composites are impermanent. Strive for perfection through heedfulness!*

This is a great address, the strongest message to the bhikkhus. In this message, the significance of the term *Tathāgata* is hinted at. Certainly, our body is a composite existence that is impermanent and changeable. One realizes, every day and every moment, that every part of our body is changing. The term *Tathāgata* suggests our potentiality of changes, or evolving nature of ourselves. In other words, like our memory, repeated practices under the Buddhist paths have great merits and will produce fruits not only in the Buddhist senses but also in the physical body of the practitioner, both internally and externally. Regarding the former sense, transformation of the practitioner's mind takes place toward attaining Buddhist enlightenment.

On the other hand, regarding the latter sense, modern sciences of the 21st century help us and give us a favorable hint, from a completely different angle. Specifically, the trace of repeated practice is enhanced in our brain during the sleep, improving connections across the brain as a whole. Let us consider a recreative and remodeling aspect of our sleep acting within the brain, physiologically and neuroscientifically.

(b) *Recreating and remodeling functions of sleeping*

The key is what happens to us during our sleep. Why do we spend so much time, as long as a third of our living time, sleeping? *Synapse plasticity* is the key word. Synaptic plasticity is the adaptation of brain neurons during the learning process.

Within the brain, transmission of impulse signals is essential along nerve fibers from one neuron to another. A synapse is a gap structure (Fig. 3.6) that permits one neuron (or a nerve cell) to pass an electrical signal to another neuron through it. Neurons are specialized to pass signals to individual target cells, and the synapse gaps are the means by which they do so. Synaptic plasticity is the adaptation ability of the connection of the synapses to strengthen or weaken the connection between them over time, in response to repeated increases or decreases in their connection activity. Since memories are represented by vastly interconnected neural circuits in the brain, synaptic plasticity is one of the important neurochemical mechanisms of learning and memory. This is called the Hebbian property, which is a neuroscientific theory, claiming an increase in synaptic efficacy arising from repeated and persistent stimulation. This explains the synaptic plasticity, i.e., the adaptation of brain neurons during the learning process [33] (Hebbian Theory). Hebb states it as follows:

> *Let us assume that the persistence or repetition of a reverberatory activity (or "trace") tends to induce lasting cellular changes that add to its stability.... When a cell A is near enough to excite a cell B and repeatedly or persistently takes part in firing it, some growth process takes place in one or both cells such that A's efficiency, as one of the cells firing B, is increased.*

The theory is often summarized as "Cells that fire together wire together."

What is important is that synaptic plasticity is activated during our sleep. Hence, it is proposed that Buddhists' repeated daily practices contribute to the restructuring of their brain neural networks. Importantly, the traces are processed during their sleep. Let us investigate its scientific aspect below.

(c) *Modern views on sleep*

Sleep is necessary for consolidation of newly acquired memory traces. Sleep is defined as a state of unconsciousness. In this state, the brain is relatively responsive to internal stimuli more than external stimuli. Sleep should be distinguished from coma. Coma is an unconscious state from which a person cannot be aroused.

Sleep affects our physical and mental health and is essential for the normal functioning of all the systems of our body, including the immune system. The effect of sleep on the immune system affects one's ability to fight disease and endure sickness. Brain activity during sleep and wakefulness is a result of different activating and inhibiting forces that are generated within the brain. Neurotransmitters (chemicals involved in nerve signaling) control whether one is asleep or awake, by acting on nerve cells (neurons) in different parts of the brain. Neurons located in the brainstem cause sleep by inhibiting the activity of other parts of the brain that keep us awake.

Sleep improves physical and mental states. In fact, sleep is a dynamic process. There are two distinct states that alternate in cycles during sleep. Each state is characterized by a different type of brain wave activity (i.e. the electrical activity detected with electrodes placed on the skull). Sleep consists of REM sleep (rapid eye movement) and non-REM sleep.

The REM sleep is characterized by rapid eye movement; during this phase the eyes move randomly and rapidly while, mysteriously enough, the body itself is fixed as if it is paralyzed. During non-REM sleep in the phase called deep sleep, all eye and muscle movement ceases. Experts believe that this stage is critical to restorative sleep, allowing for bodily recovery and growth. It may also bolster the immune system and other key bodily processes. Even though brain activity is reduced in the deep sleep, there is evidence that deep sleep contributes to thinking, creativity, and memory. We spend the most time in deep sleep during the first half of the night's sleep.

The REM sleep is usually associated with dreaming. During this phase, the eyeballs move rapidly, and the heart rate and breathing become rapid and irregular. Mysteriously enough, the muscles of the body are virtually paralyzed, while the brain is highly active during REM. This implies that something mysterious is happening in the brain in this phase, about which future study is expected to disclose in detail.

(d) *REM sleep and non-REM sleep*

Sleep happens in two stages: REM sleep and non-REM sleep. The two stages are important for functions that allow the brain and body to recuperate and develop. Parts of our brain are quite active during sleep. When we sleep, our body has a chance to rest and restore energy. However, the brain is working for 24 hours. A good sleep can help us cope with stress, solve problems or recover from daily exhaustions, and refresh ourselves with respect to how we think and feel.

There are two types of sleep stages inside the brain at night: non-REM sleep and REM sleep. REM sleep represents 20% to 25% of the total sleep time. REM sleep follows non-REM sleep and occurs four to five times during a normal 8 to 9-hour sleep period. The first REM period of the night may be less than 10 minutes in duration, while the last may exceed 60 minutes. In a normal night's sleep, bouts of REM occur every 90 minutes.

Our brain and body act differently during these different phases. Recent studies of synaptic plasticity have provided new insights into how sleep alters synaptic strength.

(e) *Synaptic plasticity*

Sleep is necessary for consolidation of newly acquired memories and synaptic traces. There is now ample evidence that sleep plays a crucial role in the consolidation of newly acquired memory, particularly for explicitly instructed practices.

Memories are thought to persist as altered connections between neurons, referred to as memories. When we practice the Buddhist path, morning and evening, over and over again, the practice strengthens the connections between associated neurons. Subsequent sleep enhances the repeated practice. Within the brain, the enhancement gets started, first by downscaling connections across the brain as a whole, thereby making free capacity for further memories, avoiding saturation, then subsequent sleep reorganizes the initial memory traces into a more robust form. Amazingly, this occurs during our sleep when the sleeper is said to be immersed in an unconscious state.

Non-REM sleep strengthens memories via two complementary processes. It suppresses the initial memory trace formed during the first exercise, and reorganizes the newly acquired information into a more stable state. The findings open up the possibility of enhancing new exercises by manipulating brain circuits during non-REM sleep.

While both non-REM sleep and REM sleep may play a role in the synaptic plasticity, the REM sleep in particular may be important for developmental plasticity. The stages of non-REM sleep and REM sleep cycle over and over again around four times during a night's sleep.

Long-term memory is consolidated from short-term to long-term memories, primarily in the hippocampus and stored throughout the cortex [A. Almarez-Espinoza and M. H. Grider, Physiology, *Long Term Memory*, 2021], called long-term synaptic potentiation (LTP). In regard to the role of sleep in the synaptic plasticity, the hippocampal LTP can be induced during REM sleep. Most studies of LTP and sleep have focused on the hippocampus.

It is known that the hippocampal place cells replay the patterns of activity during sleep, which was originally present during prior waking experience. Replay is a means of transferring information (memories) from the hippocampus to the neocortex, suggesting that replay induces adaptive, functional plastic changes in the brain. Although there is some evidence that forms of replay occur during REM sleep, communication between the hippocampus and cortex is generally conjectured to occur during non-REM sleep. Hence, the repeated cycle of two stages of non-REM and REM sleep during a night's sleep is essential for memory consolidation. This paves the way for Buddhist bhikkhus to attain the state of the Tathāgata by *diligent, uninterrupted practices.*

7.4 Strongest Last Message to Bhikkhus from Shakyamuni

Summarizing, the essence of the teaching of Shakyamuni Buddha is represented by two: "All dhammas of composites are impermanent" and "Striving to practice Buddhist Ways with heedfulness without interruption." This is a view understood from the last address of Shakyamuni Buddha of the *Maha-parinibbana Sutta* [13].

Let us see the sutra. During the last journey when the Shakyamuni and the bhikkhus took their rain retreat in the village of Beluva, the Tathāgata Shakyamuni told Ananda (Section 7.2 (b)):

> *"I am now frail, Ananda: old, elderly, advanced in years, having come to the last stage of life, now I am eighty years old." And further, "Ananda! Only when the Tathāgata is not attending to anything at all, with the cessation of feelings, and abides in the formless concentration of mind, then his body is more wholesome."*

From the awakening at Bodhgaya beneath a pippala tree at the age of 35, he has now become 80 years old. During the 45 years of a long life since then, Shakyamuni has expounded his philosophy and teachings together with uninterrupted practices. In the *Maha-parinibbana Sutta*, Shakyamuni Buddha is referring to himself as Tathāgata. The term *Tathāgata* means "one who has thus come" in Sanskrit. In the English version, the word *Tathāgata* is cited as it is, without translation into English. This means that the term *Tathāgata* has its own Buddhist sense that cannot be translated into English with a single word. We have investigated the significance of the word *Tathāgata* in Sections 7.2 and 7.3. The Japanese version translates it as "one who has come striving for the Noble Way."

Regarding the last word of *Tathāgata* in Part Six (The Passing Away), Shakyamuni addressed the bhikkhus, saying:

> *"All fabricated beings are subject to decay and impermanent. Strive for perfection through heedfulness!"*

Here, the Japanese version adds the translator's notes citing the remark of Venerable *Buddhaghosa*,[1] saying "The Buddha addressed to the Bhikkhus on the last bed ground of the *maha parinibbana* that all the teachings given for 45 years were summarized only in two simple phrases: "*All fabricated beings are subject to decay and impermanence. Strive for perfection through heedfulness.*" From this, the essence of Buddha's teaching is compactly summarized by the two: the *impermanence* and *diligent practice*. This must have been the aim of the bhikkhus authors of the scripture. However, with passage of time, Gautama Siddhartha was reasoned variously, as Shakyamuni, Buddha, Tathagata, *etc.*

(According to Waldschmidt)[2].

Shakyamuni's last message was very strong, just,

Strive for perfection of Tathagata!

---

[1] *Buddhaghosa*: [Buddha+ghosa: voice of Buddha], (flourished c.401–c.449) An Indian *Theravada* Buddhist and philosopher, who worked at Mahavihara, Anuradhapura in Sri Lanka, after the time of Vasubandhu (350–430 CE).
[2] Waldschmidt: German Indologist (1897–1985)

Standing Buddha

*One of the first representations by a
carved statue of Buddha,*

1st–2nd century CE,

Gandhāra in Pakistan.

PART II

Śūnyatā: Everything Is Empty Intrinsically

Chapter 8

No-Self, Empathy, and Great Compassion

Fig. 8.1   Walking meditation. [25]

A Neurological Evidence of No-Self

During the walking meditation (left), *mirror neurons* (colored parts) *live-report* the action of moving to oneself. The one that is reporting is not the person, but the neurons like a loud speaker connected neuro-electrically to the morter cortex. The walker may believe that there exists a self, but in reality, the neurons are telling an imitating story.

V. Ramachandran [40] (See Fig. 8.5 for mirror neurons)

8.1 Introduction, *Concerning No-self*

In the Theravāda school (Chapter 1), *śūnyatā* often refers to the non-self nature of the five aggregates of human existence and the six sense faculties and is also used to refer to a meditative state. In the Mahāyāna school, *śūnyatā* refers to an extended philosophical concept that *all things are empty of intrinsic existence* (see Chapter 9), and also refers to the Buddha-nature and the emptiness of intrinsic nature.

It is remarkable that modern brain science provides neurological evidence of *self-awareness* in the brain, which is simply a matter of neuron activity and mirror neurons. The self is not a holistic property of the entire brain (hence, not the individual as a whole); it arises from the activity of specific sets of brain neuron circuits (see Sections 8.2–8.4). Thus, modern science supports the non-self nature of the brain aggregates of human existence.

The following cites an old dispute between Buddhists, around 150 BCE, describing a point of contention in the early times: concerning *how the ātman* (soul), *or pudgala, is denied as a real entity.* The Sanskrit *pudgala* denotes a person, personal entity, body, ego, or soul.

> There are six consciousnesses: visual, ..., mental consciousness. Conditioned by both the eye and visible forms, visual consciousness arises. This visual consciousness can only cognize the visibles, not the *pudgala* [the self soul]. This *pudgala* is not cognizable by visual

consciousness; only the visibles are cognizable by visual consciousness. Thus, this visual consciousness is not the *pudgala*. [Continued further]

From the coming together of the three [the eye, visible forms, and visual consciousness], there is contact. Conditioned by the contact, there is sensation. This sensation generated by the visual contact can only sense the visibles, not the *pudgala*. This *pudgala* is not sensed by the sensation generated by visual contact; only the visibles are sensed by the sensation generated by visual contact. Thus, this sensation generated by visual contact is not a sensation generated by the *pudgala*.

([2] *AKB* (2012), Vol. I (p. 58) by Bhikkhu K. L. Dhammajoti)

It is very interesting to compare this with the fact that the triplet of visual contact (*the eye, visible forms, and visual consciousness*) corresponds to the ②, ①, and ④ of Fig. 3.2 in Chapter 3. Hence the observation of ancient Buddhists is certainly scientific from modern point of view. Summarizing the above argument:

Hence, the pudgala does not exist. As with the visual consciousness, the same is true for the cases of auditory, olfactory, gustatory, bodily and mental consciousnesses.

There was a school asserting that, while there is no *ātman* (no self-soul), there exists a *pudgala* (person) or *sattva* (being). It was in the 4th century CE, after five hundred years from the time of the above description, when Vasubandhu wrote his objection to the notion of *pudgala* (a person, personal entity, or ego) in his masterwork AKB [2], saying that "they see a self in what is not a self and fall into the abyss of views."

Vasubandhu writes: "Buddha explained that what we call a person (pudgala) is simply the aggregates," which arise depending on conditions coming together, and cites a passage from a Buddhist text as follows:

The self, in fact, does not have the nature of a self. It is through a mistaken mental construction that one thinks that a self exists; there exists no sentient being (sattva), no self (ātman), but only the dharmas produced by causes such as aggregates [Chapter 2] or the twelve members of dependent origination [Chapter 4]. When examined in depth, no person (pudgala) is perceived in them. See internal phenomena which are empty of self; See external phenomena which are empty of self, and even the practitioners who meditate on emptiness do not exist as an real entity. [2, Chap. IX, p. 2538]

In this regard, it may be interesting to cast a light on certain areas of a neural system in the brain that make the brain-owner to erroneaously assume that one's self exists. In Section 8.2, we will see the real function of the mirror neuron and possible misinterpretation of the existence of self.

According to the abhidharma theory of Section 2.4 (*vii*), "taking what is not a self to be a self" is one of the four *mistaken* views. The other three concern "what is *not permanent*, *not satisfactory*, and *not pure*" being mistaken as *"permanent, satisfactory,* and *pure."*[1]

It was right in the middle of the transitional period when Mahāyāna schools were emerging in the northern parts of India and Pakistan. The Mahāyāna Yogācāra (Yogachara) school is considered to have been founded in the 4th century by the Gandharan brothers Asaṅga and Vasubandhu. It is said in India that Mahāyāna Buddhists co-existed peacefully with non-Mahāyāna practitioners peacefully.

Ācārya Nāgārjuna (often called Bodhisattva Nāgārjuna) is credited with founding the Mahāyāna Madhyamaka school in the 1st or 2nd century CE. *Śūnyatā* was central to the Madhyamaka philosophy. Nāgārjuna's life is considered to have occurred sometime during between 75 and 125 CE. In his later years, Nāgārjuna is said to have lived in the area of Śrīparvata, located in the South India state of Andhra Pradesh. The place is now submerged in the lake Nāgārjunakoṇḍa (meaning Nāgārjuna-hill, constructed through engineering and named after the Master), which supplies water to local residents and farmers. Archeological discoveries at Amarāvatī and Nāgārjunakoṇḍa confirm the fact that Nāgārjuna maintained a friendly relationship with Sātavāhana kings.

---

[1] The four mistaken views are compactly expressed in Chinese as 常(permanent), 樂 (satisfactory), 我(self), and 净 (pure).

Three Ancient Wise Men
Bodhisattva Maitreya (*top*), Asaṅga (*middle*), and Vasubandhu (*lower right*)

Maitreya-nātha (ca. 270–350 CE), Asaṅga (ca. 330–405 CE), and Vasubandhu (ca. 350–430 CE): the three founders of the Yogācāra school of Buddhist philosophy.

The drawing of previous page depicts Asanga (left), Vasubandhu (right), and Bodhisattva Maitreya on the top with a rainbow ray of five different colors coming down. Maitreya resides in the *Tu*ṣita heaven, said to be reachable through meditation.

Asaṅga, depicted in white, is the expression of illuminated consciousness as a yogi. He engaged himself in ceaseless meditation for twelve years on a hill. He used his meditative powers (siddhis) to travel to the *Tu*ṣita heaven to receive teachings from Maitreya Bodhisattva on *emptiness*, and continued to travel to receive teachings from Maitreya on the Mahāyāna Yogācāra sutras.

On the other hand, Vasubandhu, depicted in a contrasting scholastic negligee, is pre-eminently a scholar and prefers intellectual attitude to realize the truth. However, after he encountered Asaṅga (his half-brother with the same mother), he converted to Mahāyāna in his forties and became a prolific Mahāyāna scholar of the Yogācāra school. (For Maitreya-nātha, see Wikipedia.)

8.2 Brain Neural Network and Absence of Self

When we are awake (i.e. not asleep), we are conscious and able to respond to the environment. Neural evidence indicates [36] that we first register bare sensory signals (visual, auditory, gustatory, and other general sensory signals) at the sensory cortices in the back half of the brain. As the neural impulses of the perception signals relay forward in the brain, they become much more entangled with a number of other associations. Elaborate linkages occur with language and with other sticky emotional attachments.

In the early milliseconds, as the first sensory signals register in the back of the brain, they might seem to represent the *sensing* reasonably close to reality, or at least convey an almost mirror-like sense of the object. Yet, complications lurk in every emotional overreactivity programmed into our limbic system. Each subjective veil attached to our sensory and cognitive associations obscures and distorts this reality.

The human brain is a spongy mass of dull gray nervous tissue. From such brain matter, how can our personal experience (i.e. our consciousness) arise? This is one of the most intriguing problems in neuroscience, called the *hard problem* of the mind. Usually one is conscious about oneself and becomes aware of self. To neuroscientists, the study of the sense of self-consciousness is one of the greatest challenges. All they can observe is the brain and its neural activity. *There is no structure in the brain that can be associated with a self.* Despite the apparent absence of a self in the neural activity of the brain, we have the continuous experience of a self.

The *self seems neurally absent,* but on the basis of experience, ordinary worldlings claim that it is present [35].

Our brain is made of approximately 80 billion nerve cells, called neurons. Neurons have the ability to gather and transmit electrochemical

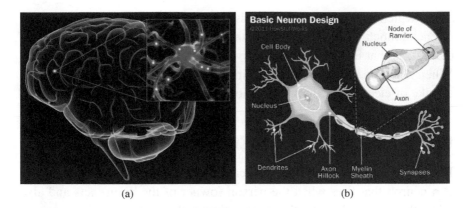

(a)   (b)

Fig. 8.2   (a) Neurotic physiology: The tissue of brain is composed of neurons. Neurons cells are electrically excitable and can receive and transmit information in the form of electric pulse waves. http://scicurious.scientopia.org/2011/05/04/science-101-the-neuron/ (b) Neuron structure: The axon hillock is a part of the cell body (or soma) of a neuron that connects to the axon. http://science.howstuffworks.com/life/inside-the-mind/human-brain/brain1.htm

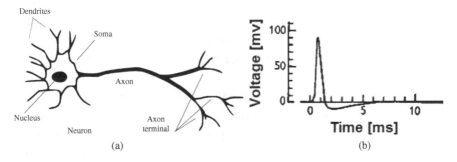

(a)   (b)

Fig. 8.3   (a) Neuron; (b) Electrical voltage signal through an axon.

signals. Neurons share the same characteristics and have the same makeup as other cells, but the electrochemical aspect lets them transmit signals over long distances (up to several feet or a few meters) and send messages to each other (Fig. 8.2).

In the brain, a typical neuron collect signals from other neurons through a host of fine structures called dendrites (Figs. 8.2 and 8.3). The neuron sends out spikes of electrical activity (Fig. 8.3(b)) through the axon, which can split into thousands of branches (terminals). At the end of each branch, a synapse converts the activity from the axon into electrical effects that inhibit or excite activity on the contacted (target) neuron. When a neuron receives excitatory input that is sufficiently large, it sends a spike of electrical activity (an action potential) down its axon.

8.3 Mirror Neurons and Empathy

In the brain, there exists a special location called the *mirror neuron*, which is a neuron that fires when the person (owner of the brain) acts and also when the person observes the same action performed by another person.

When a practitioner does an exercise such as walking meditation (Fig. 8.1), the person's mirror neuron fires. This means that when the practitioner does a *live-reporting* of the walking meditation to oneself, the mirror neuron is firing in the practitioner's brain. Hence, thanks to the mirror neuron, the practitioner is concentrating simultaneously on the exercises of both walking and mindfulness. Hence the practitioner is keeping awareness. Obviously, this is different from discursive walking. Owing to the concentration, the practitioner's discursiveness subsides after a while, and the practitioner transforms from the mere practice of *live-reporting* to an advanced level of awareness and sails into deep concentration, which is equivalent to *dhyāna*.

The continuous walk is a prolonged rhythmic exercise that causes *serotonin* to be released in the brain. Thus, the practitioners are immersed in the exercise under a pleasant feeling while overcoming discursiveness.

Next, see the photograph in Fig. 8.4. The brain neural network (of the baby) is able to *mirror* the behavior of the other person (mother) as if the observer (baby) were acting, and then the observer (baby) imitates it by action (Fig. 8.4). The story does not end there, with the mirror function of the neural network. In general, as a result of the *mirror* effect, the observer acquires another cognitive capacity, an emotional feeling such as *empathy*.

Empathy is the capacity to understand or feel what another person is experiencing, that is, the capacity to place oneself in another's position. Empathy includes cognitive empathy, emotional or affective empathy, as well as bodily and mental empathy. A mirror neuron is a neuron that fires

Fig. 8.4   An observer's imitation by mirror neurons [26].

both when a person performs an action and when the person observes the same action being performed by another. Thus, this neuron "mirrors" or reflects the behavior of the other being, as if the observer were the one acting. Such neurons have been observed in primates and other non-human species, including birds [28].

The mirror neuron was discovered by Italian neurophysiologists in the 1980s [35]. They were studying the neurons of a macaque monkey that was specialized in controlling hand and mouth movements. They placed electrodes in the premotor cortex of the monkey and found that some neurons responded when the monkey observed a person picking up a piece of food, and also when the monkey itself picked up the food.

The last point is very important in the present context, because it means that the mirror neuron in the brain is *live-reporting* one's action to oneself. It is inevitable to cause misunderstanding, that is, *the mirror neuron makes one wrongly assume that there exists a self within the brain.*

Let us consider an example. Suppose that somebody pokes my left thumb with a needle. We know that when a related cortex *a* (*insula*, Fig. 8.5) fires its cells, we experience pain. The agony of pain is probably experienced in another cortex, *c* (*anterior cingulate cortex*), which has cells that respond to pain. In subsequent pain processing, we feel pain and the affective quality of pain.

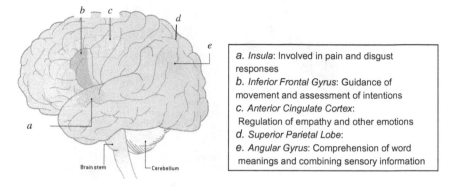

a. Insula: Involved in pain and disgust responses
b. Inferior Frontal Gyrus: Guidance of movement and assessment of intentions
c. Anterior Cingulate Cortex: Regulation of empathy and other emotions
d. Superior Parietal Lobe:
e. Angular Gyrus: Comprehension of word meanings and combining sensory information

Brainstem   Cerebellum

Fig. 8.5   Human cerebral map concerned with mirror neurons [28]

Next, when I watch a person being poked, the same neurons of *c* that responded to my thumb being poked will fire, but only a subset of *c*-cells do. In other words, there exist *non-mirror* pain-neurons as well as *mirror* pain-neurons. The latter mirror neurons are probably involved in *empathy* for pain. What the mirror neurons are doing is to allow a person *A* to empathize with the pain of another person *B*, resulting in the person *A* experiencing the same agony as the person *B*, as if somebody were poking the person *A* with a needle. That is the basis of all empathy.

Recent studies have shown that the parts of the human brain such as *inferior frontal cortex b* and *superior parietal lobe d* are active when a person performs an action and also when the person sees another individual performing that action. It has been suggested that these brain regions contain mirror neurons, and they have been defined as the *human mirror neuron system*. Possible functions of the mirror neuron [29] are as follows:

(*i*) *Automatic imitation*: An individual, having observed a body movement, performs a similar body movement;

(*ii*) *Understanding*: Mirror neurons have been postulated to have the potential to provide a mechanism for understanding actions, learning through imitation, and the simulation of other people's behavior;

(*iii*) *Empathy*: Certain brain regions are active when people experience emotions such as *disgust, happiness, sorrow*, and when they see another person experiencing an emotion.

(*iv*) *Self-awareness*: Mirror neurons may provide the neurological basis for *human self-awareness*.

Hence, when one of one's friends is in pain or sorrow, one can feel the same pain or sorrow. On this physiological basis, Buddhists practice cultivating compassion for the suffering of others. There is no doubt that the compassion of Buddhists is strengthened by the physiological function of mirror neurons. Moreover, it is a pleasant surprise to learn that the *awareness* and *consciousness* may also be based on mirror neurons [40].

8.4 Self-awareness of Mirror Neuron Network [36, 40]

It appears that the human brain is a mere lump of jelly-like matter inside our cranial vault. Despite this, the brain can contemplate and ask questions such as about the meaning of its own existence. To account for this, *mirror neurons* help us.

Neurons in the prefrontal cortex send out some signals down the spinal cord that orchestrate skilled and unskilled movements such as putting food in your mouth, pulling a lever, pushing a button, and so on. These are ordinary motor command neurons, but some of them, known as *mirror neurons*, also fire when we merely watch another person perform a similar act. It is as if the neuron and its associated network were using the visual input to do a sort of *virtual-reality simulation* of the other person's actions. This allows us to empathize with the other and view the world from the perspective of the other.

It is of profound significance to realize that these neurons have additional deep functions concerning our consciousness. The *mirror neurons* can not only help simulate other people's behavior, but can be turned to inward consciousness — as it were — to create second-order representation of our own earlier brain processes. This could be the *neural basis of introspection* and of the *reciprocity* of self-awareness and other's awareness. According to V. S. Ramachandran [40], "self-awareness" is simply using mirror neurons for "looking at myself as if someone else is look at me." The self is not a holistic property of the entire brain; it arises from the activity of specific sets of interlinked brain circuits. In this regard, see Fig. 8.1, on walking meditation, at the top page of this chapter.

Here it must be remarked that the kind of self-awareness mentioned above is connected with the function of mirror neurons, and hence that self-awareness is not said to be the evidence of the existence of *pudgala* (intrinsic self) in the sense of the abhidharma philosophy. As mentioned in Section 8.1, Vasubandhu cites a passage of a sutra, "Buddha explained that what we call a person (*pudgala*) [a self] is simply the aggregates," which arise depending on conditions coming together. The function of the *mirror neuron system* mentioned above is one of the aggregates, and thus *devoid of intrinsic existence.* The existence of a self is a mistaken mental construction. There exists no self (*pudgala* or *ātman*), but only the neuro-chemical phenomena produced by cause and effect.

8.5 Empathy and Great Compassion

The term *empathy* is the capacity to understand or feel what another person is experiencing, that is, the capacity to place oneself in another's position. The empathy described so far is also called *somatic empathy*, which is a physical reaction, based on the mirror neuron network of the somatic nervous system (i.e. the nervous system relating to body cells). The word *empathy* is derived from the Greek word *empatheia*, made up of *em*, meaning "in, at," and *pathos*, referring to "passion, feeling, or emotion." Hence, *empathy* is "physical affection or passion," including emotional empathy and somatic empathy. On the other hand, the word *compassion* is made up of the Latin words *com*, meaning "together" and *pati*, meaning "to suffer, feel pity."

In Buddhist schools, the basic motivation of all practitioners is *compassion* for the suffering of others, including all sentient beings. Undoubtedly, the compassion of Buddhists is strengthened by the physiological empathy associated with the function of mirror neurons. From the beginning, practitioners are encouraged in everyday life to practice the following Threefold Meritorious Deeds: giving (*dāna*), morality (*sīla*), and meditation (*bhāvanā*). It is often said that, of these three, the third excels, but it must be emphasized that the third deed, mental cultivation, is fragile without firm support of the first two in the background.

In the historical development from abhidharma to Mahāyāna, there has been an increasing tendency to stress on the positive attributes of practices. In the Mahāyāna development, this becomes most conspicuous, particularly in the positive concept of Buddhahood, *nirvāṇa*, and absolute reality, and the *four immeasurables* are often highlighted in the practice.

In Mahāyāna, compassion (*karuṇa*) is extended to Great Compassion (*mahākaruṇa*). In the former compassion (*karuṇa*), phenomena are considered as unsatisfactoriness derived from suffering, whereas the latter envisions the threefold unsatisfactoriness (Section 5.4). In ancient times, the three were: (i) the unsatisfactoriness of painful sensation, (ii) unsatisfactoriness upon decay, and (iii) the unsatisfactoriness attributed to all conditioned factors of existence due to their *impermanence*. All defiled composite dharmas in the mundane world are a cause for suffering and unsatisfactory because of their impermanence.

Compassion is sympathy for the suffering of others without any hatred, while Great Compassion is *non-delusion*. Delusion (*moha*) is belief *in something that is not true*, or something *that a person believes and wants to be true when it is actually not true* (from the Buddhist point of view of Four Noble Truths). From delusion, doubt arises, and individuals are confused about the truths; from doubt, false views arise; from false views, biased views arise, and so on. In these aspects, the Great Compassion of non-delusion is compared with plain compassion.

Great Compassion (*mahākaruṇa*) is equally concerned with the happiness and benefit of all sentient beings, and no other compassion surpasses its excellence. Great Compassion considers the defiled composite dharmas (phenomena) in the mundane world as unsatisfactory because those phenomena are (i) a cause for suffering, (ii) impermanent, and (iii) impure due to defiled proclivities, whereas plain compassion considers phenomena as a cause for suffering and unsatisfactory.

Compassion sympathizes with only those sentient beings who are suffering. Great Compassion pours sympathy on all sentient beings impartially. As discussed in Section 8.6, Mahāyāna is based principally upon the path of bodhisattvas, who are motivated by Great Compassion.

8.6 Bodhisattva: The Six *Pāramitās* and Four Immeasurables

A *bodhisattva* is one who is on the path striving towards Buddhahood particularly for the benefit of all sentient beings. A bodhisattva has resolved to practice the path for the relief of all sentient beings and their utmost liberation (*nirvāṇa*), and resolved to be content with the suffering of staying in the samsara in order to help them attain liberation (*nirvāṇa*). Etymologically, the word *bodhisattva* is made up of *bodhi*, meaning "awakened," and *sattva*, referring to "being, reality."

In particular, bodhisattvas have resolved to follow the six perfections (*pāramitā*), reserving their own attainment until all sentient beings attain *nirvāṇa* (Six perfections are *giving, morality, patience, effort, concentration*, and *wisdom*).

The earliest Buddhists practiced the Noble Eightfold Path (Section 5.8). The Eightfold Path is grouped into three noble overarching practices: noble conduct (*adhiśīla*), meditation (*samādhi*), and wisdom (*prajñā*). In Section 5.8, the three are interpreted by three parts of a revolving wheel representing the noble eightfold paths.

In later development of Buddhist schools, six *pāramitās* ("perfections," or "transcendents") have been highlighted as the path for the bodhisattvas of Mahāyāna.

The six *pāramitās* are:

⟡ *Dāna pāramitā* (布施波羅蜜):   generosity, giving of oneself;
⟡ *Śīla pāramitā* (持戒波羅蜜):   morality, discipline, proper conduct;

⟡ *Kṣānti pāramitā* (忍辱波羅蜜): patience, tolerance, endurance;
⟡ *Vīrya pāramitā* (精進波羅蜜):   energy, diligence, vigor, effort;
⟡ *Dhyāna pāramitā* (禪定波羅蜜): concentration, meditation;
⟡ *Prajñā pāramitā* (般若波羅蜜): insight, wisdom.

Guided by the noble path of the six *pāramitās* and motivated by Great Compassion for all sentient beings, bodhisattvas are resolved to make every effort to perform practical activities for the benefit of the whole world, reserving their own attainment until all sentient beings are relieved from suffering.

The noble paths that bodhisattvas take are dynamic, vast, and beyond logic, and are called the *four immeasurables*; in Sanskrit, the *four apramaṇa* (also called *brahmavihāra*):

(1) *Maitrī* (慈 cí, loving-kindness);
(2) *Mahakaruṇā* (悲 bēi, great compassion);
(3) *Muditā* (喜 xǐ, sympathetic joy);
(4) *Upekṣā* (捨 shě, equanimity, non-attachment).

The principal focus of the *four immeasurables* lies in the fact that a bodhisattva takes away the suffering of all sentient beings and gives them peace of mind, and feels delight in others' joy. The last immeasurable, *upekṣā* (non-attachment), is essential because a bodhisattva does not have any attachment to actions or wishes, because attachment to pleasure or other sensory feelings is a latent defilement or proclivity that must be abandoned. The last point is important in the *śūnyatā* philosophy because everything is empty intrinsically. This is considered in Chapter 9.

To be more precise, some remarks should be made about the last *upekṣā*, which is translated as *equanimity* in English. In the present context, *upekṣā* denotes "non-attachment" or "abandoning" (or "捨 shě" in Chinese translation), rather than "even-mindedness" or "indifference." The reason is as follows.

The four immeasurables (*apramāṇa*) are the super-sublime path of bodhisattvas. In the present context, the Sanskrit *pramāṇa* means geometrical measure, proof, or perception. Hence, *apramāṇa*, negating the *pramāṇa*, means immeasurable, or beyond logic. The *four apramāṇa*s are counteractions or antidotes to (1) *malice*, (2) *harmfulness*, (3) *non-delight*, and (4) *attachment* or *indifference*, which are measurable in some sense. Then, the *four apramāṇa*s are (1) unbounded *kindness*, (2) unbounded *great compassion*, (3) unbounded sympathetic *delight*, and (4) unbounded, calm *abandonment*.

The first of the four immeasurables (*maitrī*) represents wishes directed to others (all sentient beings), while the second and third (*karuṇā* and *muditā*) are empathetic mercy and joy, respectively — i.e. feeling delight with the happiness that others radiate. The last immeasurable, *upekṣā*, is *non-attachment* to those feelings and unbounded abandonment, moving beyond them. Thus, the wheel of the four immeasurables revolves indefinitely.

In terms of the intention to benefit others, the four *immeasurables* excel as a refreshing source of happiness for others. The four *immeasurables* as well as their fruits are firm and difficult to destroy. Hence they are singled out as being *punya* (merit), a power that accumulates one after the other. In meditative practices, mental states are systematically cultivated, thereby eliminating defilements such as malice, harmfulness, discontent, and attachment to pleasure and craving.

The first immeasurable, *maitrī*, in Sanskrit means friendliness, amity, good will, loving-kindness, and active interest in others by wishing for their happiness. It is the first of the four *brahmavihāra*s (divine abodes).

The second, *karuṇā*, in Sanskrit means compassion or sympathy. Great Compassion was highly praised by the Buddha more than any other virtue. It represents feeling the distress or pain of others as if it were one's own. There is an episode showing Gautama Buddha's compassion. Once, looking for a serial killer, he went into the forest because he had compassion for both his potential victims and the murderer himself. The third, *muditā*, in Sanskrit means empathetic joy, i.e. the satisfaction one experiences when others are happy. The fourth, *upekṣā*, in Sanskrit means equanimity, i.e. the detachment from pleasure and separation from one's family — hence non-attachment and abandonment.

What is pāramitā: According to widely accepted etymology, *pāramitā* is divided into *pāra* and *mitā*, with *pāra* meaning "beyond" or "the far bank, or shore," and *mitā* referring to "that which has gone," or *ita* meaning "that which goes." Thus, *pāramitā* means "that which has transcended to the other side." This reading is consistent with the Chinese expression "到彼岸" (*gone to the other side*) and also with the Tibetan translation "*pha rol tu phyin pa*."

It appears that the fifth, *dhyāna pāramitā*, of Mahāyāna almost coincides with the seventh limb, *dhyāna*, of the eightfold path of yoga (see Section 8.7) and in addition that the first four *pāramitā*s share almost common paths with the six limbs, up to the sixth limb, except "contemplation on God Isvara" (*īśvara praṇidhāna*) of the second limb *niyama*, because any concept of the Divine is out of the scope of Buddhism. In particular, the sixth Mahāyāna path, *prajñā pāramitā*, is conceptually different from the eighth path of yoga, *samādhi*. The latter yoga path is interpreted in English as *union with the Divine*, whereas the former path of Mahāyāna Buddhism is interpreted as *insight* or *wisdom*, which is the base of practical realization of Buddhism and Buddhist actions.

The immeasurables are often explained in connection with meditation (*dhyāna*), because it is considered to be a virtue based on concentration (*samādhi*). Firstly, meditation attracts the immeasurables; secondly, the immeasurables are the best qualities produced by meditation; and thirdly, meditation and the immeasurables attract each other. The *apramaṇas* (four immeasurables) belong to the third type and are considered to be an antidote to the basic defilements of hatred and craving, which refer to the universal unsatisfactoriness, painfulness, and suffering of mundane life.

Buddhist practitioners exercise to realize the four immeasurables (*apramana*) *Brahmavihara* refers to *maitrī*, *mahākaruṇā*, *muditā*, and *upekṣā*, or abandonment, whether it is in meditation or in other concentrative practices.

8.7 Yogi: Its Eight Limbs

In the 5th and 6th centuries BCE in ancient India, there were ascetic and Śramaṇa movements, and yoga practices were developed in systematic ways. Yoga, which originated in ancient India, consists of practices of physical, mental, and spiritual exercises aimed at controlling (*yoking*) and pacifying the mind. These aspects are shared commonly by Buddhist practices as well, as already mentioned in Chapter 6 (Section 6.3).

This section is presented in order to show similarity and dis-similarity of the practices and philosphies between Buddhists (Bodhisattvas, Tathaggata) and Yogis,[2] according to [31] and [32]. The term *Yogi* takes place of the *Bodhisattva* (§ 8.6) in the commentary *Prasannapada* (clear

---

[2] *Yogi*: One who is free of the afflictions of everyday existence, who sees the truth of things, and who is on the middle way.

words) by Candrakirti on NMK (to be given in Chap. 9). This implies that Yogi and Bodhisattva were used interchageably at the times of *Candrakirti* (c.600–650). However, the *śūnyatā* philosophy, to be expounded in the next Chapter 9, makes a sharp distinction between the two.

From the viewpoint of yoga schools, the immeasurables (*maîtri, karuṇā, muditā*, and *upekṣā*) are simply different aspects of universal sympathy, which remove all perversities in our nature and unite us with our fellow-men. It is true love that removes all boundaries, discrimination, and prejudice, leading to sublime unity in which there is no self and no other.

The practice of yoga is an art and a science aiming to unify body and mind. Its objective is to assist practitioners in using breath and the body to foster awareness of themselves as intimately connected to the unified whole of body-mind. In short, it is to achieve and maintain equanimity so as to live in peace, good health, and harmony with the greater whole.

Breathing technique (*praṇāyāma*) is very important in yoga. It goes hand in hand with the pose (*āsana*). In the Yoga Sutra (compiled prior to 400 CE by Patanjali), the practices of *praṇāyāma* and *āsana* are considered to be the highest form of purification and self-discipline for the mind and the body. The practices produce the actual physical sensation of heat (called *tapas*) or the inner fire of purification. According to Patanjali, this heat is part of the process of purifying the *naḍī*s (subtle nerve channels) of the body. This allows for a more healthy state to be experienced and allows the mind to become more calm. Yoga is concerned with balancing the flow of vital forces and directing them inward to the chakra system (the energy center: Chakra = wheel).

An important concept of the yoga practice is *pariṇāma*, which is a Sanskrit term describing *transformation* or change, on both a philosophical and practical level. *Pariṇāma* was already taken up in Chapter 6 in the context of *dhyāna* (Sections 6.2 and 6.3). It is one of the most important ideas in *The Yoga Sutras of Patanjali* and is described as the transformation which takes place when one is leaving a period of suffering or dukha. *Pariṇāma* is transformation not only of the mind but also of the body and

---

Tathagata: A term commonly reserved for the historical Buddha, denoting a perfectly realized one, or the one of perfect attainment,

senses. According to the yoga schools of Hinduism, *pariṇāma* means the cause of creation of the universe changing constantly. It always exists in physical matter (*prakriti*), which changes, and is responsible for creation, preservation, and destruction.[3]

In the Yoga Sutras, the eightfold path [31, 32] is called *ashtanga*, which literally means "eight limbs" (*ashta* = eight, *anga* = limb). These eight limbs serve as guidelines on how to live a meaningful life. They are a prescription for moral conduct and self-discipline; they direct attention toward one's health and help us acknowledge the spiritual aspects of our nature.

Eight Limbs of Yoga:
In brief, the eight limbs, or steps to yoga, are as follows: 1. Yama (universal morality); 2. Niyama (personal observances); 3. Āsanas (body postures); 4. Praṇāyāma (breathing exercises and control of prana); 5. Pratyāhāra (control of the senses); 6. Dharaṇā (concentration and cultivating inner perceptual awareness); 7. Dhyāna (devotion, meditation on the Divine); 8. Samādhi (union with the Divine).

1. **Yama** (universal morality): The first limb, *yama*, is to conduct universal practices that relate best to what we know as the Golden Rule: "Do unto others as you would have them do unto you." We must consider what we say, how we say it, and in what way it could affect others. The five yamas are compassion or harmlessness (*ahiṃsā*); truthfulness or honest (*satya*); nonstealing (*asteya*); sense control (*brahmacharya*); noncovetousness (*aparigraha*).

2. **Niyama** (personal observances): The second limb, *niyama*, has to do with self-discipline and personal observances. Compared with yamas, niyamas are more intimate and personal. They refer to the attitude we adopt toward ourselves. Niyama practices include developing personal meditation practices and making a habit of taking contemplative walks alone. The five niyamas are purity or cleanliness (*śaucha*); contentment (*saṃtoṣa*); disciplined use of our energy (*tapas*);

---

[3] From Yogapedia: https://www.yogapedia.com/definition/5722/

self-study (*svādhyāya*); and contemplation on God, or Isvara (*īśvara praṇidhāna*).

3. **Āsana** (body postures): The third, *āsana*, is posture. *Āsana* means "staying" or "abiding" in Sanskrit. This is a practice to calm the mind and move into the inner essence of being. As one practices the *āsana*, it fosters a quieting of the mind, thus it becomes both a preparation for meditation and a meditation sufficient of itself. One can develop inner strength and physiological flow within the body. This brings about a profound grounding spirituality in the body. The physicality of the yoga postures becomes a vehicle to expand the consciousness that pervades our every aspect of our body. The key to fostering the awareness and consciousness begins with the control of breath. In the yogic view, the body is a temple of spirit, the care of which is of utmost importance. Through the practice of *āsana*s, we develop the habit of discipline and the ability to concentrate in meditation.

4. **Praṇāyāma** (breath control): The *praṇāyāma* is to control the energy (*prana*) within the body, in order to restore and maintain health. It controls the respiratory process of in-breathing, out-breathing and then perfects relaxation and balance of body. Then the practitioner recognizes the connection between the breath, the mind, and the emotions. In the yoga, the practices of *praṇāyāma* and *āsana* are considered to be the highest form of purification and self-discipline of the mind and body. This allows a healthful state to be experienced and allows the mind to become more calm. As the yogi follows the proper rhythmic patterns of slow deep-breathing, the practices strengthen the respiratory system, soothe the nervous system and reduce craving.

5. **Pratyāhāra** (Control of the senses): The term *pratyāhāra* implies withdrawal of the senses from attachment to external objects. The word *āhara* means *nourishment*; *pratyāhāra* translates as *to withdraw oneself (retreat) and then nourishes the senses*. The practice of *pratyāhāra* provides us with an opportunity to step back and take a look at ourselves. *Pratyāhāra* occurs almost automatically when we meditate because we are absorbed in the meditation. Precisely because the mind is so focused, the senses follow it naturally. The senses become extraordinarily sharp.

Under normal circumstances the senses become our masters rather than being our servants. The senses entice us to develop cravings for all sorts of things. In the *pratyāhāra*, the opposite occurs. When we have to eat, we eat, but not because we have a craving for food.

6. **Dhāraṇā** (Concentration and cultivating inner perceptual awareness): *Dhāraṇā* means "immovable concentration of the mind." The essential idea is to hold the concentration or focus of attention in one direction. Having relieved ourselves of outside distractions by *pratyāhāra*, we can now deal with the distractions of the mind itself. A yoga authority states that the objective is to achieve the mental state where the mind, the intellect, and the ego are *all restrained and all these faculties are offered to the Lord for His use and in His service.* Here there is no feeling of *I* and *mine*. Now we can unleash the great potential for inner healing.

7. **Dhyāna** (Meditation or contemplation): *Dhyāna* is the uninterrupted flow of concentration. While the *dhāraṇā* practices one-pointed attention, the *dhyāna* is ultimately a state of being keenly aware without focus. At this stage, the mind has been quieted, and in the stillness it does not produces any thought particular at all. As we fine-tune our concentration and become more aware of the nature of reality, we perceive, "the only reality is the universal self, or God, which is veiled by Maya (the illusory power)."

8. **Samādhi** (Union with the Divine). The final limb in the eightfold path of yoga is *samādhi*. It means "to bring together, to merge." In the state of *samādhi*, the body and senses are at rest, yet the faculty of mind is in the state of awareness. The meditator comes to realize a profound connection to the Divine, an *interconnectedness* (a synonym is *interdependence*) with all living beings. With this realization comes the *peace that passeth all understanding*; the experience of being at one with the Universe. What Patanjali (in his classical Yoga Sutra) has described as the completion of the yogic path is what all human beings aspire to: *peace*, i.e. the ultimate stage of yoga — enlightenment.

(from [31] and [32])

Chapter 9

Nāgārjuna's Śūnyatā

Fig. 9.1 Photo by the author, 2014

Buddha statue standing on an island
in the man-made lake Nāgārjuna-sāgar
(formerly a historical hill, Nāgārjuna-konda, which became an island),
in Guntur district, Andhra Pradesh, India

The Fundamental Wisdom Verse of Middle Way, Madhyamaka-kārikā, is
the philosophical masterpiece of the Great Buddhist Ācārya Nāgārjuna.
According to the Zen tradition, Ācārya Nāgārjuna was the 14th master in
the lineage of the *Dharma-Lamp of Śākyamuni Buddha.*[1]

The main topic of *Mādhyamaka-kārikā* is *emptiness*. The two terms
emptiness and *middle way* are synonymous in this text. All phenomena

---

[1]Vasubandhu (c. 350–430 CE), the author of *Abhidharmakośa-bhāṣya* (considered in
Chapter 2), was the 21st Master in the lineage from Śākyamuni Buddha, with Mahākāśyapa
being the first Master.

in this world are characterized by the term *emptiness*, and the nature of reality is beyond concept. The noble Nāgārjuna helps us capture the nature of phenomena through logical reasoning by about 450 verses, each consisting of four lines. To capture the nature of reality, most logical reasonings of this text negate real existence of things and conclude that things do not exist intrinsically. This leads some people to think that this is *nihilistic*. Paradoxically, this is a complete affirmation of the phenomena in this world from the deepest point of view. Its reasoning will be provided in due course below.

In the background of this *kārikā* lies implicitly the philosophy of abhidharma and the practice of dhyāna. *Liberation* from attachment and *liberation* from any conceptual fabrication is essential to understanding the heart of the *Madhyamaka-kārikā*. *Madhya* is a Sanskrit word meaning "middle." With respect to *madhyama*, the suffix -*ma* suffix is a superlative, meaning "mid-most" or "medium." The suffix -*ka* is used to form adjectives. The word *kārikā* means "concise statement in verse." Thus, *Madhyamaka-kārikā* is translated as "Fundamental Wisdom in Verse on the Middle Way." In a Buddhist context, *kārikā* refers to the "middle path." Madhyamaka in general refers to the philosophical school associated with Nāgārjuna and his commentators.

Ācārya Nāgārjuna laid the philosophical foundation for all the Mahāyāna schools. His whole life included the years c. 75–125 CE. Many commentaries were written on the masterpiece *Madhyamaka-kārikā*. Giving representative examples, the earliest Chinese translation is by Kumārajīva (c.350–c.410) with his teacher Vimalākṣa (called Blue Eyes) [44]. Later commentary by the Indian scholar-monk Candrakīrti (ca. 600–650) is given by the translations [45, 46, 47, 48].

Madhyamaka-kārikā is divided into 27 chapters, consisting of about 450 verses in Kumārajīva's translation, or in Candrakīrti's commentary *Prasannapada* (Clear Words).

9.1 Introduction

Shakyamuni Buddha expounded what is called the Middle Way, avoiding two extremes. This is interpreted by citing a stringed musical instrument, *sitar*, and how to adjust its tone: "The tone of the sitar is controlled by adjusting the string tension." In order to play the instrument in good condition, the string tension must be adjusted such that the string is "not too tight, not too loose." This is a moderate path that is away from one extreme of self-mortification and away from another extreme of self-indulgence.

After several hundred years since the time of Śākyamuni Buddha, a deep philosophical concept was proposed by Ācārya Nāgārjuna, founding the philosophy of *śūnyatā*, or the Middle Way. This masterpiece is referred to as the "Treatise of the Middle" (*Madhyamaka-kārikā*), or Verses on the Principle of the Middle Way (*Mūla-madhyamaka-kārikā*). Let us call this treatise as Nāgārjuna's *Madhyamaka-kārikā*, or *NMK* in short. *NMK* was a revolutionary work in the sense that it had an epoch-making impact on the Buddhist schools during Nāgārjuna's lifetime as well as later generations. Owing to the deep philosophy and great popularity of *NMK*, Buddhist societies gained vitality and dynamism and also unprecedented power.

Central to the Madhyamaka philosophy is *śūnyatā*, which refers to the *emptiness* of inherent existence. All phenomena are empty of "solid entity" or "inherent existence," because they arise depending on one another. At a conventional level, "things" do exist, but in the ultimate sense they are "empty" of inherent existence. This "emptiness" itself is also "empty": it does not have an existence on its own, nor does it refer to a transcendental reality beyond phenomenal reality. The *śūnyatā* philosophy was a key philosophy in Mahāyāna Buddhism. The term *śūnyatā* (Sanskrit; Pāli: *suññatā*) is translated into English as *emptiness* or *voidness*.

In this chapter, we consider the concept of *śūnyatā* from Nāgārjuna's text *NMK* and also from the viewpoint of modern science. In Theravada schools, *suññatā* often refers to the non-self nature of the aggregates and the six sense spheres. *Suññatā* is also used to refer to a meditative state. In Mahāyāna schools, *śūnyatā* refers to the concept that *all things are empty of intrinsic existence and nature*.

Originally, the notion of *dependent arising* was referred to as the five aggregates and the *twelve links of interdependent origination*. However, Ācārya Nāgārjuna has generalized the notion of *dependent arising* that every dharma, or every "thing," does not exist on its own (hence, *emptiness*). Since every dharma depends on other "things," it is designated only provisionally or tentatively. That is nothing but what the Middle Way means. A verse in Chapter XXIV reads:

> *Whatever is dependently arising,*
> *That is explained to be emptiness.*
> *That is provisionally designated.*
> *That is itself the Middle Way.*
>
> [Verse 18, Chap. XXIV ("Four Noble Truths")]

The two terms *emptiness* and *middle way* are synonyms in the text *Madhyamaka-kārikā* (see below). All phenomena in this world are characterized by the term *emptiness*, and the nature of reality is beyond any concept. The noble Nāgārjuna helps us capture the nature of phenomena through logical reasoning with about 450 verses consisting of four lines. To capture the nature of reality, most logical reasonings of this text negate the real existence of things and conclude that things do not exist in actuality, i.e. things are of the nature of *emptiness*. This leads some people to think that the reasoning is *nihilistic*. Paradoxically speaking, *emptiness* is a complete affirmation of the phenomena in this world from the deepest point of view. Its reasoning will be provided in due course below.

In the background of this *kārikā* (verses) lies implicitly the philosophy of abhidharma and the practice of dhyāna. *Liberation* from attachment and *freedom* from any conceptual fabrication is essential to understanding the heart of *NMK*. To capture the living world truly, understanding through literal meaning (i.e. just through the brain) is not enough; instead, practicing *dhyāna* while exercising calm abiding and mindful insight is essential. For that, keeping the lotus pose (*padmāsana*) is most helpful. In practice, this enables the whole mind-body to experience the world (see Chapters 6 and 8).

About the treatise NMK

The masterpiece of Ācārya Nāgārjuna is referred to as the "Treatise of the Middle Way" (*Madhyamaka-kārikā*), or Verses on the Principles of the

Middle Way (*Mūla-madhyamaka-kārikā*); in short, *NMK*.[2] The work *NMK* is composed of about 27 chapters that explain deep Buddhist philosophy. Its central concept is *śūnyatā* (*emptiness*). Because of great popularity of *NMK*, Buddhist societies gained vitality and dynamism and also unprecedented power.

9.2   Ācārya Nāgārjuna

(*a*) *Nāgārjuna's life*
Ācārya Nāgārjuna is credited with founding the Madhyamaka school of Mahāyāna Buddhism along with his disciple Āryadeva. Although his life is not known much conclusively, Ācārya Nāgārjuna is said to have lived sometime during the last quarter of the first century CE and the first quarter of the second century CE. Thus, probably his whole life included the years of c. 75–125 CE (but other sources note the years c. 150–250 CE for his life). In the Zen School of Japan, Ācārya Nāgārjuna is respected as the fourteenth Master of the lineage descending from Mahākassapa, the first Master who had received the Dharma from Shakyamuni Buddha. In some Mahāyāna traditions, Nāgārjuna is regarded as the second Buddha.

It is said that he was born into a Brahmin family in central India, Vidarbha, in Maharashtra, and later became a Buddhist. The biography in Chinese (translated by Kumārajīva) describes an episode that led to his conversion to Buddhism. When he was young, he learned all the Vedic texts. However, out of lust of heart, he and three close friends later infiltrated the royal court. Using the art of invisibility, he narrowly escaped the

---

[2] Nāgārjuna's *Madhyamaka-kārikā*: It appears that the whole of this *kārikā* (verses) had no formal title, or its original title is not known since the original Sanskrit version was lost. An early translation of the treatise into Chinese by Kumārajīva (c. 350–410 CE) had the title "中論" [44], which is translated into English as "Treatise of the Middle." This is used in the Japanese edition, and its Sanskrit title might be *Madhyamaka-kārikā*. Later, Candrakirti (c. 600–650) studied this treatise exhaustively and claimed that Nāgārjuna's exposition was *prasannapada* (clear-worded or lucid). Hence, it was titled *Mūlamadhyamaka-vrtti-prasannapada* [45, 46, 47]. It has a Tibetan version [48] as well. The Sanskrit *mula* means "base"," "prime"," or "king's original."

hands of guards and was the only survivor. He learned that "pain is caused by desire and lust" and that "they are sources of trouble," and decided to become a monk.

Having renounced the world to become a Buddhist monk, he spent some time at the monastery that evolved into Nalanda University in Magadha, modern-day Bihar, and studied all of the *tripiṭaka*: sūtra, vinaya, and abhidharma. Then he entered the Himalayan range of mountains. It is said that Ācārya Nāgārjuna procured some *Prajñāpāramitā Sutras* from the Nāgas. The term Nāgas is understood to indicate a band of great yogis. The sutras were Mahāyāna scriptures in early historic times. To expound them, he developed a profound philosophy based on critical arguments and rigorous logics of the *reduction and absurdum* (proof by contradiction), because he thought that it was necessary to defend their Mahāyāna philosophy against attacks from opponent schools. His original philosophy is regarded as a fundamental achievement, central to understanding what the concept of *śūnyatā* (emptiness) means.

In his later years, Nāgārjuna is said to have lived in the mountain area of Śrīparvata, in the southern state of Andhra Pradesh, now submerged in the man-made lake (reservoir) Nāgārjunakoṇḍa, named after Nāgārjuna-Hill (supplying irrigation water to local districts along with electricity generation). Archeological discoveries at Amarāvatī and Nāgārjunakoṇḍa confirm the fact that Nāgārjuna maintained friendly relations with Sātavāhana kings.

According to the "Nāgārjuna's Letter to King Gautamiputra" [42], he addressed this two-part letter to King Gautamiputra. The first part describes the accumulation of merits: moral discipline, study, and meditation. The second describes the accumulation of wisdom, i.e. realization of Lord Buddha's profound wisdom on the basis of *interdependent origination*, leading to Buddha's enlightenment. In the letter of Ratnavali, Nāgārjuna instructed the king to chant a Dharma verse three times a day in front of an *image of the Buddha* and to construct images of the Buddha *positioned on a lotus*. The reign of King Gautamiputra is dated variously: 86–110 CE, or 106–130 CE. However, according to Ref. [43], the most likely kings of Nāgārjuna's patron are those of a later generation, from c. 175 to c. 220 CE. Hence, Nāgārjuna' lifetime varies depending on the literature.

(*b*) *Nāgārjuna's philosophy*

Nāgārjuna is well known as the author of *Madhyamika-kārikā*, which is one of approximately fifteen works attributed to him. The real achievement of the masterpiece is based on *rigorous* logical analysis of phenomena by revealing incoherence of the mental idea that things possess an *intrinsic* nature (entity) — i.e. an eternal nature that does not depend on anything external. Moreover, the idea of *self* is our own conceptual, delusory fabrication. If we stick to the idea that there is a *self* that has an unchanging innate *entity*, it should result in inconsistence. To prove this, Nāgārjuna systematically used logical arguments, i.e. *proof by contradiction* (i.e. *reductio ad absurdum*[3]).

All phenomena arise together in a mutually interdependent web of interactions. The emergence of everything depends on something else. Nothing exists independently. Based on this observation, he attacked, in his verses, the concept of "entity" or "intrinsic nature" *(svabhāva)*. Stanza 18 of Chapter XXIV reads:

> Whatever is dependently co-arisen,
> That is explained to be *emptiness*.
> Its existence is imputed in dependence upon something else,
> This is the path of the Middle Way.

What is the Middle Way?

> "Everything exists": That is one extreme.
> "Everything doesn't exist at all": That is another extreme.
> Avoiding these two extremes, the Tathāgata taught the Dhamma via the middle. This is the Middle Way expounded by the Tathāgata.

Nāgārjuna's ultimate principle of emptiness was equated by him with "dependent co-arising," "dependent origination," or "interdependent

---

[3] Probably, Nāgārjuna was one of the earliest persons in history to use this logical *proof by contradiction*. Suppose that P is the proposition to be proved. The *reductio ad absurdum* is as follows: (i) P is assumed to be false, i.e. ¬P (non-P) is true. (ii) It is shown that ¬P implies two mutually contradictory assertions, Q and ¬Q. (iii) Since Q and ¬ Q cannot both be true, the assumption that P is false must be wrong, and P must be true.

arising." This is expressed by the Sanskrit term *pratītyasamutpāda*, which means *arising in relation to causal conditions*, or arising on the basis of causal conditions.

According to this philosophical view, there is no distinction between *nirvāṇa* and *saṃsāra* (transmigration; cyclic wandering among worldly existences). This will be considered later again. Understanding by intelligent logical words only is not enough for capturing the real living world. Dhyāna practice is its essential base of Nāgārjuna's philosophy. The *Commentary to Nargarjuna's Yuktisastika* by Candrakīrti (600–c.650) describes the state of *dhyāna* as follows:

> *"From peace of mind, the body becomes most clean. With clean body, one enjoys peace of mind. From the mental peace, mental concentration is achieved. From the concentrated mental state, proper perception is realized and proper understanding is achieved. Proper recognition and correct understanding help detachment from worldly affairs. By the detachment from worldly affairs, attachment to worldly affairs diminishes. Free from the attachment leads to moksha, liberation from samsara."*

9.3 Verses on the Principles of the Middle Way (NMK)

Mūla-madhyamaka-kārikā is translated as *Verses on the Principles of Middle Way*. Three central concepts of this masterpiece would be (*i*) *pratītyasamutpāda* (dependent origination or dependent co-origination), which is a Buddhist view on the world; (*ii*) *śūnyatā* (emptiness), which is the wisdom of Bodhisattva and the essence of the path *prajñāpāramitā*; (*iii*) *nirvāṇa* (liberation from suffering, called *extinguished*), which is the mindful and peaceful path of Buddhists.

The deep analysis of this masterpiece is supported by a combination of two personal approaches: (*a*) unprejudiced observation of our real world and (*b*) strict logical reasoning, aided by extensive Buddhist literature he had access to.

In Section 9.3, the verses of *NMK* are interpreted based on two esteemed classical commentaries, which have been kept and remained in

China, Japan, and Tibet for centuries. They have been translated into Japanese and English in the last forty years. The classical commentaries are (i) the earliest version of the commentary on *NMK* by Master Vimalākṣa (also called Blue Eyes) translated into Chinese by Kumārajīva (c. 409 CE) [44], and (ii) the Sanskrit (and Tibetan) version of the commentary *Prasannapada* by Candrakīrti (c. 650 CE) [45, 46, 47, 48], where [48] is described mainly by the author's own comments.

(1) *What is the dependent origination*
In the first chapter of *NMK*, the key to understanding the entire *Verses on the Principles of the Middle Way* is presented, where detailed interpretation on the concept of *dependent origination* is developed. Let us start by considering it with examples.

(a) First, let us consider the *five aggregates* presented in Section 2.5. Observing our daily life and ourselves at every moment, we experience a continuous change in the state of ourselves and the external phenomena around us. In Section 3.2, we learned that "seeing an object is the perception represented by a chain sequence of signal transductions within our body. The visual system is regarded as a stimulus-response system." Seeing is *visual consciousness, which is an aggregate of sensations*. All the five aggregates (*forms, sensations, conceptions, formations,* and *consciousnesses*) are regarded as dependent origination and viewed as a stream that has *twelve links of interdependent origination*.

(b) Second, *dependent origination* was expounded first by the Tathāgata, Śākyamuni Buddha. This is recorded in the sutra "The Great Elephant Footprint Simile" (*Mahāhatthipadopama-sutta*) [49], a chapter in *Majjhima Nikāya* (Collection of Middle-Length Discourses):

The Tathāgata has said, "Whoever sees dependent co-arising sees the Dhamma; whoever sees the Dhamma sees dependent origination." And these things — the five clinging-aggregates — are dependently co-arisen. Any desire, embracing, grasping, & holding-on to these five clinging-aggregates is the origination of stress. Any subduing of desire & passion, any abandoning of desire & passion for these five clinging-aggregates is the cessation of stress [duḥkha]. And even to this extent, friends, the monk has accomplished a great deal.

(translation by Thanissaro Bhikkhu, 2003)

(c) Third, an intelligible example is provided by a biological phenomenon in nature, i.e. a plant sprouting from a seed:

A sprout of a seed, being the sprout, is not the seed itself; hence a sprout arising from a seed is not generated from itself; nor is it wholly other than the seed; yet it is not the same; it does not arise spontaneously, it needs some scientific conditions to be satisfied; it does not arise from God. In other words, the plant object is neither perishable (it does not stop being) nor permanent (it does not keep the same).

It is said: "Whatever arises (comes into existence) dependent on something else cannot be that very same thing; nor can it be wholly different other either; thus it neither stays unchanged permanently nor ceases to exist completely." Let us consider further examples.

(d) A pot is made from clay in cooperation with the supporting conditions of water and fire-heat. According to conventional reasoning, a pot is *effected* from the *cause* of clay by *scientific conditions* of water and heat. It does not make sense to consider that a pot (as an effect) pre-existed in the material of clay (as a cause). Thus, a thing does not arise from itself. Moreover, before the pot is made, it does not make sense to consider that the clay and water are causal conditions for the pot because those would be conditions for non-existence (because the pot is not there). In addition, one may say that if the pot is heated to a temperature much higher than that at which it was made, it shall melt into a liquid state or enter a gaseous state at even higher temperatures, and eventually diffuse into space. This implies that the pot, supposed as an effect, has no *entity* and is not considered as an inherent existence or self-existence.

(e) Remembering the previous example (c) of plant sprouting, the sprout does not pre-exist within the plant seed. In other words, such an object (considered as an effect) does not pre-exist in the conditions (considered to be causal). It is noted that the seed and its sprout do not coexist simultaneously, but appear sequentially. Thus, the sprout is not identical to the seed. But it is not the otherness of the seed either.

Subsequently, from the sprout grow leaves, stalks, buds, and flowers consecutively. The flowers produce seeds. This cycle continues repeatedly and endlessly. It is an endless sequence of a plant's transformation from one state to the next. Within the sequence of plant states, one can hardly find an inherent existence that does not change.

(f) It is said that a cloth consists of threads and the cloth is made from an *efficient* cause of threads. If the threads themselves were self-existent, i.e. if they were a real entity, the cloth would exist as an entity consisting only of threads. Here, *entity* is conceived as an inherent existence or self-existence. Consider cotton threads. They are made of a number of cotton fibers with fiber walls consisting of innumerable spirally oriented cellulose fibrils, and so on. Hence, threads are composites and not self-existent. Although the cloth is viewed as an effect made of threads, it is composed of many sub-parts (sub-structures) that have no self-nature and are not self-existent.

Cloth is supposed to be realized from a cause, and this cause again from another cause.
How can what is not realized in its own right be the cause of something else?
Things which thus arise as effects are not self-existent.

[*Nāgārjuna*]

(2) *Opening theme: pratītyasamutpāda*

(a) Opening of *NMK* (Nāgārjuna's *Madhyamaka-kārikā*):
Ācārya Nāgārjuna explains the concept of dependent origination as follows:

Whatever arises in dependence on something else does not arise in truth (in real meaning).

This is termed *pratītyasamutpāda* in Sanskrit, which can be described by the following Eight Negations. At the very outset of *NMK*, Nāgārjuna presents a verse of dedication:

I prostrate to the perfectly enlightened one, Buddha Śākyamuni,
the best of teachers, who expounded the pratītyasamutpāda as

Whatever is dependently arisen is
Neither ceasing to exist, nor arising,
Neither terminable, nor permanent,
Neither self-identical, nor distinctive.
Neither coming, nor going,
And free from conceptual fabrication.

[*Eight Negations*: Dedicatory Verse of *NMK*]

When we examine a group of conditions that give rise to a thing or an entity (such as a pot, or a sprout, or cloth), any analysis of the causal conditions does not show the fact that such an object exists among the conditions. Moreover, the conditions themselves are not self-existent, and the things that arise as effects from those are not self-existent. This is termed *dependent origination*, and described by the author Nāgārjuna as *pratītyasamutpāda* in Sanskrit, stating that everything arises depending upon multiple causes and conditions and that nothing exists as an independent entity.

Nāgārjuna's primary intention is to expound the nature of dependent origination. He has developed this treatise *NMK* out of compassion and for the enlightenment of others.

(*b*) First verse, Chapter I:
The great treatise *NMK* begins with the following verse:

No things exist whatsoever,
having arisen of themselves,
from another, from both or without cause,
at any time, in any place.

(Verse 1, Chap. I)

The first assertion is as follows: "Not, as arisen of themselves, do any things at all exist, ever, anywhere at all." The other three assertions are stated in the same way: "Not as arisen from another, do any things at all exist, ever, anywhere at all."

Buddhapalita (an earlier commentator, c. 500 CE) says, "Things do not arise of themselves because such origination would be purposeless and because it entails an absurdity. There would be no purpose in the repeated origination of things which are in existence already."

The basic stance of Nāgārjuna's logical analysis was described by himself in his later work, *Refutation of Objections*:

"If I were to advance any thesis with a point at issue whatsoever, that in itself would be a fault; but I advance no thesis and so cannot be faulted."

"If, through the means of valid knowledge, I cognized any object at all, I would affirm or deny its existence; but as I do not do this, I am not culpable."

(*c*) Consciousness cannot be captured conceptually
The *very present moment* (the mental present) of ourselves is self-awareness and nothing but *consciousness*. One can ask, "Is there unmediated self-awareness, or else?" Let us see the quotation from the sutra "Questions of Ratnacuda" [45, Chap. II, p. 55]:

The Bodhisattva[4] contemplates the mind and enquires into the stream of consciousness, asking "whence does the consciousness arise?" And he thinks: "Consciousness arises given an object exists"," with the object mediating the awareness. Does this mean that the object is one thing and consciousness another? Or are they identical? If the object is one thing and consciousness another, there will be a duplication of consciousness. If they are identical, how can one perceive consciousness by means of consciousness? But, the consciousness does not perceive consciousness. This is because the edge of a sword cannot cut its own edge, nor can the tip of a finger touch that very tip. In the same way, one act of consciousness cannot directly perceive the same act of consciousness.

---

[4] One on the path to achieve *Samyaksambodhi* (perfect enlightenment), motivated by great compassion.

If consciousness could know itself, one cannot deny that the edge of a sword can cut its own edge, a tip of a finger can touch that very tip, and an eye can see itself. However, see the continuation, from the passage of the sutra:

> So the Bodhisattva, concerned with what arises in the mind, observes that the mind is transforming without beginning nor ending. The mind is not unchanging, is not uncaused, nor unconditioned, is neither identical with itself nor different. The mind is groundless. He knows and sees the stream of consciousness like a twining creeper plant. He knows and see the essential nature of consciousness: the groundlessness of consciousness, the hiddenness of consciousness, the imperceptibleness of consciousness, and the absolute uniqueness of consciousness. As he knows and sees it thus. So, he knows it and sees it as it really is, and he does not suppress it.
>
> This is the analysis of consciousness as he truly knows it and sees it. This is the Bodhisattva's contemplation of consciousness, and this is his sacred penetration into thought through thought.

Thus, consciousness cannot directly perceive the same consciousness. Consciousness cannot be captured conceptually. This is just what is the reality of our world.

9.4 First chapter: "Enquiry into the Conditions"

The first chapter of *NMK* presents the fundamental key concept and logical structure to understand the entire *NMK*, where not only a sequence of logical reasonings are developed, but also detailed explanations are given on fundamental concepts and definitions resulting in the main theme, *dependent origination*. The title of the first chapter is "Enquiry into the Conditions," and the whole chapter is devoted to investigating what is the nature of a *condition*.

What is a condition? All phenomena in this world are in Sanskrit called *saṃskṛta*, translated as *conditioned phenomena*,[5] which is

---

[5] Conditioned phenomena (saṃskṛta, 有為) are *what have been made*, a synonym of *sasrava* (impure phenomena), connected with defilements or afflictions, with the exception of the noble path of the enlightened one [2: Chaps. 1 & 2].

explained to mean *what has been made* or *put together* by causes (*hetu*) and conditions (*pratyaya*). This is expressed by a single term, *pratītyasamutpāda*, as explained in the previous subsection. The *abhidharma* doctrine [2: chapters I & II] categorizes the *conditions for arising* of phenomena into four classes, which are explained in item (*f*) ("*Conditions*") given below.

Investigating the *conditions for the arising* of phenomena, the author Nāgārjuna presents rigorous logical analysis of key concepts such as "cause", "effect", "condition of objective basis", "non-condition", "things with self-existence", "inherent existence (or entity)", "non-existence", and so on. Let us consider these key concepts one by one. The first verse of the first chapter is already presented in Section 9.3(2)(b). After extensive logical reasoning, it is finally concluded that the notion of *conditions* is refutable, and that an *effect* is *non-existent*. Thus, we will arrive at a point where we have a clearer view of the conditioned phenomena described by *pratītyasamutpāda*.

(a) ***Non-condition***:
Visual perception (an effect) arises depending on the eye, visual material objects, etc., as its conditions. These are said to be its *efficient* conditions.

> *These give rise to those effects,*
> *so these are called conditions.*
> *As long as nothing arises,*
> *why are these not non-conditions?*

(Verse 5, Chap. I)

Comment on the last two lines: As long as visual perception (the effect) has not arisen, it is undeniable that the eye, material objects, etc., are non-conditions.[6] The author Nāgārjuna is seriously concerned with whether a *condition of a perception object* makes sense when perceptive consciousness is absent. See the following.

---

[6] There can be no arising from non-conditions. Even sand grains are a non-condition for sesame oil [45, 47, 48].

(b) *Condition of objective basis*:

A thing is said to exist *inherently* if it exists by virtue of possessing an *entity*, for it exists independently of other entities. Taking an example of visual consciousness, the eye and visual forms (objects) are conceived as the conditions of *perceptual consciousness*, or *conditions of objective basis*. At one time, those are conceived as the conditions of an existing visual perception, while at another time those are not conceived as objects of visual perception. In the latter case, visual perception is not existent with respect to those visual forms. If something is non-existent, how can there be a *condition* at all for what does not exist? The sixth verse of Chapter I is as follows:

> *A condition makes no sense,*
> *either the effect is existent, or non-existent.*
> *How could there be a condition for the non-existent?*
> *How can a condition apply to what is existent?*

To comment on the last line, what is existent means that it exists inherently (already). Why is a condition for producing it (or its arising) necessary?

This is a proof by contradiction, i.e. by *reductio ad absurdum*. Suppose that visual perception *exists* as an effect. It should exist inherently by definition. Hence, any condition for producing it (for its arising) is unnecessary. If visual perception is *non-existent*, the non-existence is not brought forth because the visual perception is not in existence.

In conventional reasoning, a condition brings forth an effect. The counterproof by contradiction (i.e. *reductio ad absurdum*) is as follows. The statement P to be proved[7] is that *a condition does not make sense*.

If P is not true, *an effect* is brought forth from *a condition*. If the effect is existent inherently, any condition for producing it is unnecessary, because it exists inherently by definition. If it is *non-existent*, the non-existence is not brought forth by any condition. Accordingly, any effect is not brought forth from the condition. Hence, whether the effect exists or not, the notion of *a condition makes no sense,*

---

[7]P is defined *logically* in the footnote in the subsection "Nāgārjuna's philosophy" of Section 8.2(*b*).

Thus, the notion of the *condition of objective basis*[8] is refuted. In fact, the objective-basis condition is conceived (*fabricated*) to explain the arising mental content, which is supposed to exist in the *delusive* everyday world. If mental content is without an objective basis, it is factually non-existent.

(c) *Things without self-existence*:

A thing that lacks self-existence is never a real entity, where *entity* is meant as an inherent existence or self-existence. If a thing is not self-existent, how can it be represented as a *cause*? And if something arises from what is not self-existent, how can the arising thing be represented as an *effect*?

Someone may question. Having seen that clothes are made of threads, it is said that threads are the condition of the clothes. A piece of clothing does not exist within any of the conditions such as the threads, the loom, and so on, which are considered to lack self-existence. Cloth is not perceived in the conditions considered as causal. If the person insists further, "Even if an effect is not existent in the condition, it can arise from conditions." Nāgārjuna replies,

[A]: *Why should an effect not issue forth from non-conditions?*

(Verse 12, Chap. I)

Logically speaking, restating [A], "Is it reasonable to negate the statement that an effect arises from non-conditions?" Hence, the whole statement [A] implies that an effect with self-entity does not arise from conditions without self-existence. Thus,

[B]: *the conditions in general are inefficacious because of their inability to give rise to effects.*
So, no effect is produced, and there is in consequence no cause.

---

[8] The item (*f*) ("*Conditions*") lists four conditions: (i) condition of cause, (ii) condition of objective basis, (iii) immediately preceding condition, and (iv) condition of dominance (or efficient cause).

(d) *Immediately preceding condition*:
Sprouting from a plant seed is viewed differently from the previous example of cloth made of threads (condition (*iv*), see (f)). According to the abhidharma philosophy, the growth of a plant from its seed is conceived as a continuous sequence of states at different moments, and each existence (state) at an instant perishes before the next arises; as it were, each phase is supposed to have power to bring forth its successor.

Suppose that the seed has ceased to exist before the effect arises. If this is so, when the seed has ceased to exist, it is non-existent. What will be the cause of the next sprout phase? What will be the cause of the extinction of the seed?

If something has indeed come to an end, how will it be a condition?

The seed and other factors (temperature, water, sunlight, etc.) are considered to have ceased to exist when the sprout does not come into existence. If this is true, it must be that both the extinction of the seed and the arising of the sprout are without cause. Thus, the immediately preceding condition does not make sense, and the notion of *immediately preceding condition* is refuted.

However, modern science investigates a number of natural phenomena resembling the phenomena represented by the *immediately preceding condition*. In fact, the continuous sequence of states of a plant is called *evolution* or *transformation*. Those are of utmost importance from the point of view of science. This is considered in the item "Transformation in science," given below.

(e) *Effect is non-existent*:
The author Nāgārjuna summarizes Chapter I, titled "Enquiry into the Conditions," with the following verse:

> *The effect is supposed to be caused from conditions,*
> *The conditions do not have their own self-entity,*
> *If the effect arises from conditions without self-entity,*
> *How could an effect come from conditions?*

[Verse 13]

The last line implies that things with self-entity would not arise from conditions without self-entity. Cloth is supposed to be realized from a causal condition. But how can what is not self-existent be the cause of something else? Things (as an effect) would exist if conditions existed, or non-conditions existed. If the effect existed, we could say these are its conditions or those are its non-conditions. That is not the case.

(f) *Conditions*:

From the point of view of *abhidharma* philosophy [2: Chaps. I & II], let us consider the notion of *conditions* for arising. So far, we have seen that things do not arise from themselves, nor do they arise spontaneously. Things do not arise from what is other things. Even though we saw this, the idea of condition is commonly accepted, and the condition as *cause* (*hetu*) is defined by the property that it brings forth something.

All phenomena in this world are in Sanskrit called *saṃskṛta*, translated as *conditioned phenomena*. This is explained to mean that which *has been made* or *put together* by causes (*hetu*) and conditions (*pratyaya*). The second verse of Chapter I ("Enquiry into the Conditions") summarizes it from the abhidharma doctrine:

> *The conditions of arising are four: (i) condition of cause, (ii) condition*
> *of objective basis*
> *(iii) immediately preceding condition, and (iv) condition of dominance*
> *(or efficient cause).*
> *There is no fifth condition.*

The first, *condition of cause*, is defined as what actualizes something else. Secondly, when mind-content (*dharma*) arises, it is supposed to arise with reference to an object. This defines the second condition of *objective basis*. Thirdly, the extinction of an immediately preceding factor is assumed as the condition for arising of the effect. The fourth condition is the factor because of whose existence something else comes to be. This defines the *condition of dominance*, or *efficient* cause.

The second, *condition of objective basis*, has been considered already in detail in item (*b*), where the notion of the *condition of objective basis* is refuted. In the delusive mundane world, the concept of objective cause is fabricated (created) to explain arising mental flows as if those are boiling or firing. However, consciousness and its content exist factually without objective causes. If mental content is without an objective basis, the objective cause itself cannot exist. Hence the hypothetic definition of the *condition of objective basis* has no reality.

The third, *immediately preceding condition*, is also refuted. It is considered in item (*d*). If it is true, it must be that both the extinction of the preceding factor and the arising of the subsequent factor are without cause. That does not make sense. Thus, the hypothetic definition of the *immediately preceding condition* does not have reality either.

It is unreasonable to consider a condition for something that does not exist. To consider such a condition is impossible. If something is non-existent, how can there be a condition for that thing? Consider how to define the *horn of a rabbit* that does not exist. This does not make sense.[9] This example is provided to demonstrate whether it is reasonable to define a non-existent object.

Let us now examine the remaining conditions, (*i*) *condition of cause* and (*iv*) *condition of efficient cause*, and consider whether those definitions are reasonable. The definition of condition as *cause* (*hetu*) is that the condition brings forth something, i.e. produces it.

Suppose that an element of existence (dharma) is to be brought forth. If it were really brought forth, the producing cause would produce it as an effect. But in reality, that is not the case. This is already investigated in the items "(*c*) *Things without self-existence*" and "(*e*) *Effect is non-existent*"."

An inherently existing thing (an element of existence) is independent by definition, and it cannot be produced by anything else. An inherently non-existing thing cannot be produced either: if it were there, it would be existent. How could one propose a *condition of cause* to bring it forth? It does not make sense.

If *this* exists, then *that* arises. This is the *assumed* definition of *efficient cause*. But nothing is self-existent; all things arise in mutual

---

[9] However, one may define the horn of cattle.

dependence. How can the "this" be represented as a cause? And how can "that" be represented as an effect? Therefore, the condition of *efficient cause* is not established. Thus, the author Nāgārjuna summarizes Chapter I ("Enquiry into the Conditions") in the last verse:

> *Therefore, an effect is made existent (does exist)*
> *Neither with conditions, nor with non-conditions*
> *As the effect is non-existent,*
> *How could there be either conditions or non-conditions?*

[Verse 14]

After all, Ācārya Nāgārjuna refutes the notion of either *conditions* or *non-conditions*.

9.5 Selected Verses from NMK

The whole body of *NMK* (Nāgārjuna's *Madhyamaka-kārikā*) is composed of about 450 verses, with the total number verying slightly depending on translation versions. The entire system of the verses develops a lot of brilliant ideas and impressive statements and presents poetic and logical expressions of the Middle Way, from which some representative verses are selected below.

(1) *The eye cannot see itself*

The eye is a visual sense organ, and what the eye sees is said to be a visual object. Vision is the act of seeing an object, and visual perception is represented by a sequence of signal transductions occurring in the human internal sense organs (Section 3.2, Fig. 3.2). The visual system is regarded as a *stimulus-response* system.

Let us pose a question. What does vision see? Verse 2 of Chapter III declares,

> *Vision does not see itself.*
> *How can something which does not see itself see other things?*
> [Verse 2, Chap. III (Enquiry into the Eye and Other Sense Faculties)]

The first line of the verse is analogous to the statement "consciousness cannot directly perceive the same consciousness," considered in item (*c*) ("Questions of Ratnacuda") of Section 9.3(2). In fact, vision is visual perception. The point is not that vision is impossible, but that visual consciousness is unable to be captured conceptually or described by words. If one dares to put it in words, one must say that the edge of a sword cannot cut its own edge. *Visual consciousness* is the *very present moment* of ourselves. According to the Bodhisattva, the essential nature of visual consciousness is the imperceptibility of consciousness. Visual perception is a sequence of signal transductions occurring in the human sense organs and brain, stimulated by incoming rays of light. The visual system is a *stimulus-response* system (Section 3.2). Therefore, visual perception is a dependent origination that is inherently void.

The second line is concerned with another issue. This is interpreted as follows. Concerning *NMK*, there exist two classical commentaries.[10] Both of them describe the following argument of an opponent: "Although vision cannot see itself, it can see other things. So vision will see other things, but not itself." Visual perception arises as a result of being stimulated by the incoming light-rays from other things. The nature of the dependent origination is emptiness, as is the nature of vision itself.

The agent of burning is not burned. Scientifically speaking, the burning (firing) agent is a chemical reaction that occurs at high temperatures between the burning object and the oxygen in the air. Hence the agent of burning is a chemical reaction that is not burned. This is similar to visual perception, which was interpreted above as a sequence of signal transductions, equivalently the *stimulus-response* reaction, taking place electrically along visual neural elements of the sense organs. Just as the visual perception of signal transduction does not see itself, the burning agent of chemical reaction is not burned.

Author's comment:

> "The eye does not see itself. How can something which does not see itself see another?"

---

[10] An early version by Vimarakusha (also called Piṅgala) was translated into Chinese by Kumārajīva (c. 409 CE.) [44], and the Sanskrit/Tibetan version is the *Prasannapada* by Candrakīrti (c.650 CE) [45].

This verse is likely to induce laughter. In other words, "One cannot see one's own fault, but can see another's fault." Correctly speaking,

"One does not see one's own fault. How can someone who does not see one's own fault see another's?"

(2) *Dependent origination is inherently of nirvāṇa nature*

Dependent origination is the mother of this world. Observing ourselves at every instant and in our daily lives, our state and external phenomena around us are changing continuously, and all things are formed through causes and conditions that come together and coexist in assemblage, and are called "composite" (*saṃskṛta*, 有為, constructed) or sometimes "conditioned" in the Buddhist view of the world. The nature of world phenomena is characterized by the three phases of arising, abiding/transforming, and ceasing. However, from the point of view of Ācārya Nāgārjuna, things appear to be transforming selflessly, harmlessly, and peacefully. Their inherent nature is nirvāṇa. All phenomena appear to be interdependent and empty.

The Tathāgata said, "Whoever sees dependent origination sees the Dhamma; whoever sees the Dhamma sees dependent origination." Thus, verse 16 of Chapter VII states,

Whatever is arisen dependently.
Such a thing is inherently of the nature of nirvana.
Therefore what is arising is in nirvana.
And the arising itself is so.
[Chap. VII (Arising, Abiding, Ceasing)]

In the Sanskrit version, the term *nirvāṇa* is not used, but the term *sant* is used in place of *nirvāṇa*. *Santa* has various meanings, such as "peace of mind," "tranquility," "subsided," or "purified." In the Chinese version, however, the translator Kumārajīva (c. 409 CE, [44]) used the term *nirvāṇa*. Nirvāṇa has similar meanings in English: "blown out," "extinguished," "quieted," or "perfect calm."

In summary, things are not produced by causes and are neither in being nor not in being.

(3) *Is there reality in arising and ceasing*?

Our perception and mind are continuously arising and ceasing. All phenomena (i.e. all dharmas) in the mundane world are explained as composites (*saṃskṛta*, 有為). They are formed through (presumed) causes and conditions coexisting in assemblage (hence they are *composite*). Arising, transforming, and ceasing are considered the three phases appearing in our mundane world. However, those do not have real entities at all. What appears does not really exist, but is an illusion and like a dream. The three phases are mere appearance. Thus,

> *Arising, transforming and ceasing are*
> *Like a dream, like an illusion,*
> *Like a city of Gandharvas.*
> *So have the three phases been expounded:*
>   [Verse 34, Chap. VII (Arising, Transforming and Ceasing)]

Full discernment of *dharma*s is essential in attaining liberation from suffering. The phenomena of arising, transforming, and ceasing do not exist inherently and lack the real entity of existence. Similarly, the composites (or things that are conditioned) do not exist inherently and lack self-existence.

Thus, what appears in this mundane world is like a dream. However, what is said to be *like a dream* or *like an illusion* is the existence of things conceived in the mundane world, where they are considered to exist inherently. This *existence* can be given a meaning only in a conventional, relative sense. As far as one can analyze, one finds only dependence, relativity, and emptiness. Considering that phenomena have the kind of existence that is a dependent arising, one can capture the existence of phenomenal reality in the context of emptiness [48].

With the above verse, the first part (Chaps. I to VII) of *NMK* is brought to a close [48]. The treatise *NMK* (Nāgārjuna's *Madhyamaka-kārikā*) consists of twenty-seven chapters. According to Jay L. Garfield [48], *NMK* is divided into four sections as follows:

- First section (chapters I through VII): This section examines Buddhist perspectives of the mundane world such as cause and condition, sense faculties, aggregates, primary elements, and composites.

- Second section (chapters VIII through XIII): This section focuses on self-nature, suffering, and subjective experience.
- Third section (chapters XIV through XXI): This section is concerned with the self, entity, actions, and fruits.
- Fourth section (chapters XXII through XXVII): The final section addresses the ultimate truth such as the Tathāgata, emptiness, nirvāṇa, Four Noble Truths, and the twelve links.

(4) *Cessation of conception: leading to the path of pacified goodness of nirvāṇa*

Some persons view things falsely as existing inherently because their mind is obscured by a network of various conceptual thoughts of worldly ignorance. This is as if a big log of wood were covered with a network of creeping plants. Those thoughts fall away from the true path of seeing things unerringly. The path leading to *nirvāṇa* is subtle. Thus,

> *Those who see existence and non-existence of things*
> *Are of inferior mind.*
> *They do not capture what is seen*
> *As pacified goodness of nirvana.*
>
> [Verse 8, Chap. V (Primal Elements)]

Following this verse, the Chinese version [44] (translated by Kumārajīva c. 409) adds the following commentary given by his teacher Vimalākṣa (called Blue Eyes):

Wise-men annihilate the view of *existence* when they see ceasing of phenomena (things). Wise-men annihilate the view of *non-existence* when they see arising of phenomena (things). Thus, although there are various views with respect to every phenomenon (dharma) in this world, it is like an illusion, or like a dream. Therefore, if one does not see pacified nature of the *view-annihilation*, the one gets the sight of *existence* and gets the sight of *non-existence*.

Does it make sense to conceive the *property* or *characteristic* as existing *substantially* (*independently*)? Or, are they *interdependent* and therefore *empty*? This raises the question of whether *conception* has a realistic meaning. *Conception* is one of the five aggregates considered in Section 2.5 (*iii*).

It is commonly accepted that everything has a characteristic. For example, consider a male deer with a distinguishing characteristic, such as its big horn. Analogously, the distinguishing characteristic of *space* is its *openness* [45] (*Absence of material forms* [44]).[11] But can the characteristic exist inherently in space? How do characteristics come to be associated with space? Prior to its characteristic *openness*, space could not be characterized because it could not have been without *openness* characterizing it.

If space existed prior to its distinguishing characteristic *openness*, it would follow that it was without this characteristic, and space without its distinguishing characteristic *openness* would have been actualized. It is like a *deer* without *horns*. It does not exist. Thus, a thing without a characteristic cannot be actual.

Consider a donkey, which has no horn. Any distinguishing characteristic (horn) cannot be actualized in a donkey (lacking a horn). In the other case, newly actualizing a distinguishing characteristic (horns) to something (cattle) that is already so characterized is unintelligible. It is like actualizing horns in cattle, which already have horns. The horn actualization is not realistic on the donkey (which lacks it) and on the cattle (which have it) as well. An opponent may say that the subject of characterization does exist because the characteristic (horn) is a fact.

The donkey and cattle horn examples have made it clear that distinguishing characteristics cannot be realized on either the donkey or cattle, which are subjects for actualizing horns. In other words, nothing can exist separately from both characteristic and what is characterized. The characteristic and what is characterized exist interchangeably. Consider the combined pair of fire and smoke. The fire is characterized by the smoke,

---

[11] According to classical Buddhist cosmology, there exist six primal elements: *space, earth, water, fire, air* (*wind*), and *consciousness* [45, 48]. Because of this scriptural statement, someone may say, "Even as the primal elements exist, so do the basis of perception and personal existence." Chapter 5 ("Primal Elements") of *NMK* examines how things are characterized in terms of properties or characteristics.

while the smoke is characterized by the fire [44]. They exist in a *mutually interdependent way.*

Likewise, *space* is characterized by *openness* (the absence of material forms), while the openness is characterized by space. This reasoning poses the question of whether the characterization (*conception*) of a thing (*space*) and its characteristics (*openness*) have realistic meaning, or whether they exist *substantially*? Or, are they *mutually interdependent and therefore empty*? The discerning and characterizing fine distinction is *conception.*

This raises the question of whether *conception* has a realistic meaning. Conception (one of the five aggregates) is considered in Section 2.5 (*iii*). It is through conception that a *name* is given to external visual impression. In other words, conception is discerning and can make fine distinctions.

In this way, the significance of achieving the cessation of conception is understood. The attainment of the cessation of conception is called *liberation*, because it averts the conception of external objects or the emergence of all conditioned phenomena. Noble ones practice this attainment of cessation because they consider it as a peaceful abode of concentration.

Thus, Ācārya Nāgārjuna concludes as follows. Conception is discerning, which is characterized with the ability to make fine distinctions. By attaining the cessation of conception, nothing can exist separately from both characteristic and what is characterized. If these two (characterization and characteristic) do not exist on their own, there can be no *space*. The five primal elements are e*arth, water, fire, air,* and *consciousness*, which are exactly like the *space*.

Thus, *space is neither an entity nor a non-entity.* The Tathāgata said:

Whoever comprehends things as non-entity clings in no way to anything at all. Whoever clings in no way to anything at all touches the samādhi-state without discerning individual distinctions.

The entire manifest world arises from discriminative thinking. The world dharmas should be realized (understood) to be subtle and beyond the reach of thinking.

If someone thinks the elements of existence merely empty, the one is immature and walks a dangerous path. [(To think them non-empty is similarly argued as well.]

(5) *Emptiness emerges immediately when all views are cast off*

When all views have been cast off, emptiness emerges immediately, because the base of all views and phenomena is *emptiness*. Because of the *nescience* (ignorance) of the unenlightened one, the wisdom eyes are dimmed and afflicted. It is like mistaking a wheel of fire for a circling torch. A torch with a burning tip (resembling a point) is only circling in the space.[12] The wheel of fire lacks self-existence and does not exist as a real entity.

The wheel of fire is a case caused by the limited performance of the human nervous system. Generally speaking, all (composite) things or phenomena pretend to exist, although they have no real entity. They lack self-existence and they do not exist as real entities. The absence of *being* (i.e. *real entity*) in things is *śūnyatā* (*emptiness*):

Enlightened ones have said
that emptiness is the relinquishing of all views.
For whom emptiness is held as a view [holding like a real entity]
They are regarded to be incurable.

[Verse 8, Chap. XIII (Absence of Being in Things)]

Observing that everything varies constantly, an opponent may ask, "If a thing lacks an essential nature and therefore it is empty, what could change?" Change would mean the change in some attribute. But since all

---

[12] The view of a circling torch is perceived as a fire circle by human eyes. Scientifically, this is inevitable because the visual recognition of images and movement is controlled by the brain's *alpha-wave* in the visual cortex. The typical wave period of the alpha-wave is about 0.1 second. Hence, if the circling torch takes 0.1 second to complete one revolution, the visual cortex is unable to resolve the detailed position of its burning tip. Therefore, the view of a fire-wheel is not due to the intellectual ignorance of the observer. *NMK* was written about two thousand years ago. But it is true that a circle of fire lacks self-existence and does not exist as a real entity. It is caused by the ability and the inability of the brain's neural system — that is, the *nescience* (*agnostic* aspect) of the human vision system.

(composite) things are declared to lack self-existence, it is not logical to base an attribute of something on a non-existent subject, just as discussing the hardness of a donkey's non-existent horn is not logical.

Ācārya Nāgārjuna replies: If a thing (a phenomenon) exists as a real entity, it cannot vary. It is emptiness itself that makes variation comprehensible. In fact, the essential nature of weather is empty. Therefore, the weather varies daily without interruption. A characteristic property that is invariable in a thing is said to be its essential nature. Thus,

If there is an essential nature in a thing
How could it be possible to become other.
[Second half of Verse 4, Chap. XIII (Absence of Being in Things)]

Even in the case of emptiness (absence of being), some people pertinaciously hold on to the reality of empty things (phenomena), i.e. a dogmatic view of emptiness. Buddha expounded the great healing medicine *emptiness* to cure suffering people. But how can the great healer attend to those suffering the disease of holding pertinaciously on to a dogmatic emptiness? They are hopelessly lost.

In the Ratnakuta Sutra, the Tathāgata (Buddha Bhāgavat) said to Kaśyapa:

*It is not the emptiness-being which renders the phenomena (dharmas) "empty"; rather the phenomena (dharmas) are **empty by nature**. In the same way, all phenomena (dharmas) are formless by nature, and all phenomena (dharmas) are inaction (intention-less activity) by nature. Just this way of insight into phenomena, Kāśyapa!, I call it the Middle Way. The emptiness is the exhaustion of all views. It is the true way of approach to the dharmas.*

For a person who holds a dogmatic view to emptiness, no curable way is conceived. Kāśyapa!, it is as if a sick man were given a medicine by a doctor, but that the medicine, having removed his ills, was not expelled but remained in the intestines. What do you think, Kāśyapa!, will this man be freed of his sickness? No indeed, Bhagavan (Buddha), the sickness of this man in whose intestines the medicine having removed all his ills remains and is not expelled. Bhagavan said: In this sense,

> *Kāśyapa!, the emptiness is the exhaustion of all dogmatic views. But
> I call the person as one for whom the emptiness itself becomes a fixed belief.*
> [From *Prasannapada* of Candrakīrti (c. 650 CE); [45] (p. 150); [46]
> (Chap. XIII, p. 410)]

(6) *What is the Middle Way? Transcending disputation*

In the sutra "Discourse to Katyayana" (*Katyayanavavada*), Buddha said
to Katyayana:

> Asserting that things exist inherently is to fall into one extreme of reifi-
> cation. Arguing that things do not exist at all is to fall into the other
> extreme of nihilism. Following the Middle Way is neither asserting that
> things exist, nor that things do not exist.

In the sutra, Buddha claims that reification leads to grasping, craving,
and suffering, which result from failure to see impermanence [48].
Nihilism results from failure to see the phenomena of arising and ceasing,
and leads to suffering from lack of mindfulness in daily life. The Middle
Way leads to engagement in the world without attachment. It follows that
it does not make sense to insist that the true way of things can be seen in
terms of existence or non-existence. Thus,

> *Buddha refutes both views of existence*
> *and non-existence insightfully,*
> *and is detached from saying "it is" or "it is not"*
> *in the sutra Discourse to Katyayana,*
> > [Verse 7, Chap. XV (Existence and Non-existence)]

This verse is a translation from the Chinese version [44] by
Kumārajīva (c. 410), which is structured clearly to express what is meant,
more than the English versions [45, 46, 48] from Candrakīrti's
Prasannapada (c. 650).

- From the *Ratnakuta Sutra* [45] (**in-between two extremes**):
  To say "*Something is*" is one extreme; to say "*Something is not*" is one
  extreme. What are there **in-between**, avoiding these two extremes?

Those are characterized as *"being un-conceivable"*, *"unable to give specification"* *"unable to verify"*, *"without support* (or *abode)"*, *"being formless"*, or *"being signless"*. Kaśyapa! It is the Middle Way (*madhyama pratipad*); it is the right way of regarding the true nature of things.

- From the *Samadhiraja Sutra* [45] (**Avoidance of disputation**):
  "It is" and *"It is not"* are two dogmas; *"purity"* and *"impurity"* are two dogmas. The wise man abandons both dogmas without taking up a position in-between. *"It is"* and *"It is not"* are mere **disputation**; *"purity"* and *"impurity"* are mere *disputation*.

  Afflicted existence is not terminated by engaging in endless disputations. Afflicted existence is brought to an end by not engaging in disputation.

(7) *No means can annihilate karmic consequences because they do not arise inherently*

Karma means intentional action, good or bad. In fact, *karma* is a driving force of the world. *Karmic* action inevitably bears afflictions of mind. Because afflictions do not arise inherently, no means can annihilate them inherently.

The Buddhist view on our world is concerned with the *karmic* actions that cause suffering. The sequence of *karmic* actions gives rise to the *saṃsāra* (cyclic existence) of aimless wandering in the mundane world. Buddha's view is as follows:

> *Karmic action is empty by nature, however will not perish.*
> *It is of existence, but not being permanent.*
> *Both the action and effect are non-expiring.*
> *Thus expounded by the Lord Buddha.*
> [Verse 20, Chap. XVII (Examination of Karmic Action)]

All actions and phenomena are devoid of inherent existence; they are elements of the cyclic existence (saṃsāra), but will be impermanent. No effect of an action expires. In other words, the consequences of actions do not cease spontaneously. All actions have *ramifications* into the indefinite future due to continuous dependent origination [48]. Actions themselves do not inherently exist and are not a suitable basis for the inherent cessation of something that originated previously.

> Karmic action lacks inherent existence.
> Therefore, action does not arise inherently.
> Because it is not arisen inherently,
> it is not expiring inherently.
>> [Verse 21, Chap. XVII (Examination of Karmic Action)]

All phenomena are indeed empty and impermanent. All actions and consequences neither arise inherently nor cease inherently. Since all karmic actions are empty inherently and the consequence of action transcends all views, one can see the *pacified nature of nirvāṇa* in them (item (4) of this section). This resolves the apparent conflict between *karmic action* and *nirvāṇa*. The next item will give the answer.

(8) *Like an illusion, like a mirage, or like a dream*
Karmic actions and consequences are empty, and the consequence of action transcends all views. When afflicted views are stripped away, the background space of *emptiness* emerges immediately. One may say that it is existence without inherent essence, like an illusion or a dream. But that does not prevent it from being perceived. The whole sequence of an action, its consequence, and the acting agent is understood as a single empty sequence, as is the emergence of a new agent. Karmic action, its consequence, and saṃsāra are real empirical phenomena but illusory. They are empty in nature.

Just as the teacher, through divine power, emanates an illusory body, the illusory body emanates another illusory body (verse 31):

> *An agent and its action* are appearing *like that way.*
> *The agent is like the first illusory body.*
> *Its karmic action is*
> *Like the illusion's illusion.*
>> [Verse 32]

In fact, afflictions, karmic actions, and the consequences are illusory:

Afflictions, karmic actions, bodies,
Agents, and their fruits are all,
Like a city of Gandharvas and
Like a mirage or a dream.
                    [Verse 33, Chap. XVII (Examination of Karmic action)]

Mirages and dreams are actual physical and physiological phenomena, which actually appear and which actually have consequences. Those can be detected by scientific means. But that does not mean that they appear to us in a nondeceptive way [48]. The road mirage (inferior mirage) causes the observer to see the image of a bluish sky on the heated ground in the distance, created like an illusion. This is just the bright sky seen in the distance which appeared like reflection from a water puddle acting as a mirror. The road mirage is not a reflection from the surface of water puddle, but an optical refraction of light ray owing to the layered density of heated air. Water in the dream has no wetness at all. By analogy, karmic action, its consequence, and saṃsāra are real empirical phenomena like mirages, but are empty in nature.

(9) *What happens when mental fabrication ceases?*
Afflictions, karmic actions, personal existence, and the fruits of action are all deceptive and not real, but rather merely appearing to the unenlightened consciousness in the guise of reality, much like an illusion, a dream, a rainbow (optically created by rays of sunlight and cloud-droplets floating in the air), or a mirage city (similarly created as an optical image in the air). Then, on earth, what is the way things really are?

All afflictions and faults (whatsoever) arise from holding the view that the self is real. The person (yogi) practicing yoga, based on the wisdom gained from yoga practices, does not take the self as real and thus abandons such a viewpoint and stops intended karmic actions. Then all afflictions cease spontaneously and naturally:

Action and afflictions having ceased, there is liberation.
The afflictions and karmic actions arise from conceptual thought.

This comes from mental fabrication.
Fabrication is put to an end in the sphere of emptiness.
                    [Verse 5, Chap. XVIII (How Things Really Are)]

(10) *Limitation in language*

There is a fundamental limitation in languages. If there were something real for language to refer to, there could be a didactic argument (intended to instruct something). However, when what language refers to is no more existent, there is no didactic argument. If there were an object of thought, then speech would be able to function by imputing a specific character to it. However, when no object of thought exists, how can a specific character be imputed by which speech would function? The true nature of things (dharmata) neither arises nor perishes. Discursive thought cannot function with respect to it, and this hints at what nirvāṇa is in terms of written letters. However, it must be noted that nirvāṇa transcends what is expressed by languages.

When the object of thought is non-existent,
There is nothing for language to refer to,
The true nature of things
neither arises nor ceases, as Nirvana does not.
                    [Verse 7, Chap. XVIII (How Things Really Are)]

It is established beyond question that the Tathāgata has didactically argued nothing whatsoever, because the true nature of things neither arises nor perishes, and specific character cannot be imputed to things. Thus, no doctrine about anything at all has been taught by Buddha.

(11) *The way things really are*

The way things really are is independent of conceptual imputation. It is going on by itself without mediation, without others' help, and with harmony. Thus, the real way of things is

Not dependent on anything other than itself, already in nirvana,
Not fabricated by mental fabrication

Beyond thought analysis, and beyond distinction.
That is the real way things are.

[Verse 9, Chap. XVIII (How Things Really Are)]

"Not dependent on anything other than itself" is interpreted by a metaphor in the following way. Those suffering an optical defect in their eyeballs see illusory hairs, bees, and other things that do not really exist. However, once they (i.e. those with defective eyes) are cured by a wisdom medicine of seeing transparently (since things are *irrefragably* [irrefutably] without substance), they realize directly for themselves that this is the true nature of such things (nothing is there). The phrase "already in nirvana" refers to the view of the wisdom eye, which does not see illusory hairs.

(12) *Neither identical nor different: a taste of nectar with which Buddha enlightens people*

When scientists observe a continuous sequence of changing weather in daily life, they understand that the atmosphere is transforming ceaselessly. The natural sequential phenomena are *neither identical nor different*. The butterfly's life cycle is another example.

Butterflies have the typical four-stage insect life cycle. Winged adults lay eggs on the food plant, on which their larvae, known as caterpillars, feed. The caterpillars grow and, when fully developed, pupate in a chrysalis. When metamorphosis is complete, the pupal skin splits, the adult insect climbs out, and after its wings have expanded and dried, it flies off. This is why the nature of sequential phenomena is said to be neither identical nor different.

A rice sprout comes into existence because of a rice seed and the surrounding temperature, water, and other conditions. If there is a seed, there is a sprout, though the sprout is not precisely the same as the seed, nor is it totally different. Again, this shows that the nature of sequential things is neither identical nor different.

Āryadeva (3rd century CE), a disciple of Nāgārjuna, was an Ayurvedic medicine doctor-monk. He says in his work *Catuḥśataka* (Four Hundred Verses): "Seeing the fact that things do function, they are not nothing.

Seeing the fact that things cease functioning, they are not permanent." If one practices according to such enlightening medicine of viewing the world, one can make afflictions die and stop mental fabrications, and gain peaceful nirvāṇa. Thus,

> *Not identical, nor different,*
> *Not permanent, not annihilated,*
> *Such enlightening medicine of Buddha is*
> *Tasting like a nectar (amritha).*
>
> [Verse 11, Chap. XVIII (How Things Really Are)]

This enlightening medicine illuminates the profound meaning of the ultimate truth: that dependently arisen phenomena are neither inherently identical nor different, nor annihilated nor permanent.

A plant's life cycle from seed to seed formation is as follows: *seed, germination, growth,* and *flowering.* The continuous sequence is "not identical, nor different; nor annihilated, nor permanent." When those phenomena are observed, people who are suffering feel relieved, their minds are soothed, and they feel peaceful. Thus, Buddha enlightens suffering people with a taste of nectar (*amritha*).

(13) *Tathāgata: thus goes the one who has realized the Truth*

According to Pāli canons, Gautama Buddha called himself Tathāgata. The term is regarded as meaning either "one who has thus gone" (*tathā-gata*) or "one who has thus come" (*tathā-āgata*) [50]. Another *plausible* interpretation was given by a Russian Indologist, Fyodor Shcherbatskoy (according to "Tathāgata," Wikipedia). Citing from the epic Mahabharata (Shantiparva): "Just as the footprints of birds (flying) in the sky and fish (swimming) in water cannot be seen, Thus (*tātha*) is going (*gati*) of those who have realized the Truth." See item (24) below.

The Tathāgata is called so since he realizes and awakens the reality of all dharmas (things) in the way things are, not only from one's mind but also from one's whole body. The Tathāgata is not an inherently existent personal self. Why it is said that the Tathāgata is not self-existent? The first verse of Chapter XXII ("Examination of Tathāgata") reads:

Tathāgata is not identical with the aggregates,[1]
Nor different from the aggregates.[2]
The aggregates are not in the one,[3] *nor is the one in aggregates.*[4]
Tathāgata does not possess the aggregates.[5]
What on earth is the Tathāgata?

This is referred to as the *five-fold* analysis (as shown by superscript numbers). This is an analytic demonstration of non-existence of the self through analysis of the aggregates. *What on earth is the Tathāgata?* After a sequence of logical reasonings in terms of several verses following the preceding one, the last two of Chapter XXII conclude as follows:

Tathāgata is beyond mental fabrication, yet eternal like the world.
People develop cognitive fabrications with regard to Buddha.
Their wisdom eyes are harmed by their own fabrications,
And do not see the Tathāgata.

[Verse 15]

Those who spend their lives with ceaseless mental fabrications fail to see the Tathāgata because karmic actions cause their eyes to become murky. The Tathāgata is eternal and beyond all fabrications of sentient beings. Yet sentient beings are driven by karmic actions arising from basic afflictions. Then, what is the nature of the Tathāgata?

Whatever the nature of Tathāgata.
That is the nature of the world of sentient beings.
The Tathāgata is without inherent nature;
The world too is without inherent nature.

[Verse 16, Chap. XXII (Examination of Tathāgata)]

In cosmology, the cosmic world is described by physical principles expressed in mathematical forms. The cosmic world develops dynamically according to the physical laws and goes on by itself without mediation. The cosmic space has no invariant structure, and changes its structure ceaselessly. Yet, it develops according to physical principles and changes ceaselessly.

The Tathāgata is eternal. This sounds like the physical world. The Tathāgata is without inherent nature, i.e. *empty* of inherent nature. Therefore, the appearance of the Tathāgata changes ceaselessly, and it is beyond mental fabrications.

(14) *Goodness-delusion, badness-delusion, and error-delusion*

Passionate desire (*rāga*), hatred (*dveṣa*), and delusion (*moha*) are rooted in goodness-delusions, badness-delusions, or error-delusions. Love or affection is caused by a goodness-delusion, aversion is caused by a badness-delusion, and confusion is caused by an error-delusion. Those are afflictions or delusions that are rooted in self-existence and might serve as a ground for the existence-delusion of saṃsāra.

Those mental states are expressed by the term *kleśa* in Sanskrit, representing the mental states that cloud the mind and manifest in unwholesome actions. *Kleśa*s include states of mind such as anxiety, fear, anger, jealousy, desire, and depression. To represent the term *kleśa*, the terms *affliction* and *defilement* are used in English.

If afflictions and delusions were inherent existences, it is hard to see how nirvāṇa is possible because it requires the elimination of the affliction-existence. But, on the other hand, if afflictions do not exist inherently, why are there existences suffering saṃsāra? To put it another way, why are we not enlightened in such a way that we are already in nirvāṇa?

Once all the afflictions and defilements have been eradicated, nirvāṇa would emerge immediately before one's eyes. How does that happen? *Desire* is born of volitive (wishing) thought. If it is not born in thought, it does not arise anymore.

Desire, hatred, and delusion are said to be born of *nescience* (ignorance). The three afflictions are the roots of others. These three basic afflictions arise because of the goodness-delusion, badness-delusion, and error-delusion. Desire arises directly from what takes the form of goodness-belief; aversion is dependent on badness-belief; and illusion arises from misbelief.

What causes delusions or illusions? Nescience (*avidyā*) does, which is the origin of groundless acts; a confused act of consciousness is born of illusion. Thus, delusions or illusions come into being because of *avidyā* (nescience, ignorance). From a modern scientific point of view, *avidyā* would be nothing but the activity of the brain, which functions to cause passionate love, irrational aversion, or insane delusion, and also gives rise to inaccurate perceptions of rapidly moving objects (e.g. the fire wheel of a moving torch) or very tiny particles of submicron sizes.

Ācārya Nāgārjuna's insight surpasses all the thoughts and all mental fabrications:

> *If someone's afflictions*
> *Exist without entities,*
> *How could they be vanished?*
> *Who could vanish the existence without entity?*
> [Verse 25, Chap. XXIII (Afflictions and Misbeliefs)]

It is impossible to nullify existences without entity. Who can make afflictions vanish, which by their very nature do not exist? No one at all can make them vanish.

As there is no possibility of elimination, afflictions cannot be eliminated. As there is no elimination, how can there be a search for effective means of eliminating the afflictions? The *Samadhiraja Sutra* says [45, 46]:

> An affliction "desire" would be roused in someone by something; an affliction 'hatred' would be roused in someone to something; an affliction "delusion" would be roused in someone concerning something. Consider a person who does not see such afflictions nor perceive them in fact. One, who does not see such afflictions nor perceive them in fact, is said to be free of desire, hatred and delusion, to have a mind free of misbelief, and to be in *samādhi* (profound contemplation). The one is said to have completed the crossing, to have crossed to the other shore, to stay in nirvāṇa.

(15) *Conventional truth and Ultimate truth*

Whatever is born of conditions is not regarded as inherently born, because it does not come to be in self-existence. Whatever is dependent on conditions is said to be *empty* of a self-existent nature. Whoever understands the *emptiness* of self-existence is free of delusion. The meaning of *emptiness* is "being dependent on conditions," that is, being essentially unable to stand on its own.

Thus, the meaning of *dependent origination* and *emptiness of essence* is the same. But the meaning of *non-existence* is not the same as that of *emptiness of essence*, as already explained initem *(13)* concerning the Tathāgata.

It is fundamentally important to realize that we now have two opposing concepts on the same world: existence and emptiness. The former is the view of conventional truth connected with the everyday world of verbalized transactions. Consider an example. In regard to the butterfly's life cycle, it is as follows. A young larva of the insect (known as a caterpillar) feeds on a food plant. When it fully developed, it pupates in a chrysalis covered with a hard skin. When the pupal skin splits, the adult butterfly expands its wings and flies to the next life.

By contrast, the latter view (emptiness) is the view of the ultimate truth of *nirvāṇa*. The former phenomenal world is like an illusory existence, defined by names and concepts. In the latter world of the ultimate truth, phenomena are empty (*śūnyatā*) of an inherent self or essence. From the view of ultimate truth, the *phenomenal world* of the sequential nature of the butterfly's life mentioned above is viewed with eightfold negations (*emptiness*): "neither ceasing to exist nor arising; neither terminable nor permanent; neither identical nor different; neither coming nor going," and "free from conceptual fabrications."

The *phenomenal world* and *emptiness* are not two distinct things. Rather, as the above paragraph on the life of a butterfly implies, **they are two characterizations of the same thing**. It is concerned with how one can achieve liberation. Ācārya Nāgārjuna explains as follows:

> *Without a base in the conventional truth,*
> *Significance of the ultimate truth cannot be taught.*
> *Without understanding the significance of the ultimate,*
> *Liberation is not achieved.*
>
> [Verse 10, Chap. XXIV (Four Noble Truths)]

(16) *Dangers of misperception of the emptiness*

If the teaching of *emptiness* is wrongly grasped, it destroys the unwise man. If one takes the view that all things are unreal and non-existent, one is not wise and would be destroyed because of misinterpreting the world and wrongly grasping what is happening. On the other hand, if one does not deny the reality of all things, one must reject the notion of the "absence of being" in things and explain how things can be empty of self-existence. However, what *emptiness* really means is that all phenomena arise together in a mutually interdependent web of interactions and hence everything arises dependently and is not self-existent. Rejecting this view, one will inevitably proceed to calamities as a result of inappropriate deeds lacking reality. Thus,

> *By a misperception of emptiness,*
> *A person of inferior intelligence is destroyed*
> *Like a snake incorrectly seized*
> *Or like a magic ritual incorrectly applied.*
>
> [Verse 11, Chap. XXIV (Four Noble Truths)]

(17) *Dependent origination, emptiness, and the Middle Way*

Whatever is born from supporting conditions, it is not truly born because it does not arise as self-existent. Whatever depends on conditions is empty of self-nature. To say that something is empty is another way of saying that it arises dependently. For example, a chariot presupposes a cockpit, wheels, etc. The chariot composed of parts (cockpit, wheels, etc.) can exist depending on parts; therefore, it does not have self-existence.

It may be surprising to say that another aspect of any dependently arisen things is associated with *verbal conventions*. To say that something has arisen dependently is to say that its identity as a single entity is nothing more than being the referent of a *word*. To say that a thing's identity is a merely verbal fact is to say that it is empty. To view emptiness in this way is to see it neither as an entity nor as unreal. It is to say it as conventionally real but nothing more than empty [48]. It is said that "being empty'" is the absence of self-existence. A mere verbal fact (*verbal convention*) is nothing but empty.

This absence of self-existence is said to be the Middle Way. The following two verses contain the *śūnyatā* philosophy in a nutshell and are often quoted in Mahāyāna schools:

> *Whatever is dependently arisen*
> *That is interpreted to be emptiness.*
> *It is nothing more than verbal convention.*
> *It is itself the Middle Way.*
>
> [Verse18, Chap. XXIV (Four Noble Truths)]

> *There is no existence whatsoever,*
> *Which does not arise dependently;*
> *Therefore, there is nothing whatsoever*
> *Which is non-empty*
>
> [Verse 19, Chap. XXIV (Four Noble Truths)]

(18) *Impermanence and Suffering*

Whatever is impermanent is painful and causes emotional distress. It is suffering, and there is no happiness in it. Buddha said, "Impermanence is suffering." Thus,

> *If it is not dependent on conditions,*
> *How could suffering come to be?*
> *Impermanence is expounded to be suffering*
> *Because what is impermanent cannot have its own essence*
>
> [Verse 21, Chap. XXIV (Four Noble Truths)]

What is impermanent is suffering. This is the First Noble Truth.

(19) *The Four Noble Truths*

If a practitioner sees that whatever is dependently arising is of the nature of *emptiness*, he or she is ready to see the reality of the Four Noble Truths: *suffering*, *its origin*, *its cessation*, and *the path:*

Whoever sees dependent arising,
Sees suffering, origin of the suffering,
Its cessation and the path to its cessation
Namely sees the Buddha.

[Verse 40, Chap. XXIV (Four Noble Truths)]

The last line "Namely sees the Buddha" summarises the verse in a nutshell. This is given only in the kumarajivais translation [44].

When asked "How should the Four Noble Truths be seen?" Buddha replied [46]:

"Whoever sees that all things are originated *dependently*, has known thoroughly the sufferings. Whoever sees that all things have *not arisen inherently*, has already eradicated the origin of sufferings. Whoever sees that all things are ultimately *in nirvāṇa* has already realized the cessation of sufferings. Whoever sees that all things are ultimately *empty in* nature has already been practicing the noble path."

(20) *Tetra-lemma of Buddha's teaching*
The Tathāgata said: "A worldly person disputes with me. I do not dispute with the worldly person." The Tathāgata is so merciful and says people-friendly to arouse deep understanding of this world. Every person is different from others in terms of their experiences, bodily conditions, families, gender, etc. Therefore, the Tathāgata deliberately uses every clever means to guide people along the Buddhist paths to liberation. These Buddhist stratagems (Buddhist ways by which one reaches one's aim) are called *upāya* in Sanskrit, and 方便 in Chinese and Japanese. Following *tetra-lemma* is one of them:

Everything is real, and is non-real,
Both real and non-real,
Neither real nor non-real.
This is what the Lord Buddha cured people.

[Verse 8, Chap. XVIII (How Things Really Are)]

This is a flexible teaching for those who are to be led with the interests of gradual instruction, so that men and women are led from byways to the right way [45, 46].

First, what is accepted by the ordinary world is accepted by the Tathāgata; what is not accepted by the ordinary world is not accepted by the Tathāgata. Things such as aggregates or personal existence in the ordinary world are treated as *real* by the Tathāgata. This is a path of friendly mercy with an eye on the ultimate truth and with a view to arousing the faith of the ordinary man in the Tathāgata.

Second, the Tathāgata knows without a doubt what is cause, what is effect, what is pleasurable, and what is painful. Thus, the Tathāgata knows everything and sees everything as it is. Having realized the omniscience of the Tathāgata, those who suffer are told that *everything is not real* when considered deeply. This is because all compounded things change in over time. Because of this fact of change, they do not actually exist.

Third, it is explained to some that everything in the world is *both real and not real at the same time.* For those who have started on the path, everything in the world is real and everything is delusive because it is not perceived as such from a noble conscience.

Fourth, there are those who, through long-time practices, see things the way they really are and who have virtually but incompletely eradicated the obstructions to enlightenment. To them, it is explained that everything in the world is neither real nor non-real. In order to remove what remains of the obstructions, both alternatives are rejected: everything in the world is not real because it is delusive, nor non-real because it is not non-existent.

Specifically, everything is conventionally real; everything is ultimately unreal. Everything has characteristics of both the real and the unreal; everything is neither real in one sense nor unreal in another sense [48]. For example, one's suffering is conventionally real, but suffering is ultimately unreal; it is an illusion. Suffering has characteristics of both the real and the unreal. However, one's suffering is neither real because it is delusive nor unreal because one is suffering and it is not non-existent.

Language has limitations. What is expressed by language is a mental fabrication and does not really exist in the world. It is not possible for a foreigner to understand what is spoken in a language that is not one's own, because language is mentally fabricated within one's brain and separated from reality in the ultimate sense. For example, sports such as football can be played harmoniously by players speaking different languages. The tetra-lemma is the means of liberating oneself from the trap of the brain's neural network.

Such flexible tetra-lemma teachings of Buddha on the true nature of things are like nectar (*amrita,* 甘露), which cures suffering people and guides them to the ultimate path.

(21) *Nirvāṇa and conventional world*
In the everyday world, we experience the succession of birth and death, or the arising and passing away of everything. It is simply expressed as *coming and going* and may be understood as things being dependent on what is outside themselves, or originating from complex causes and conditions. The *coming and going* coincides with the subjects studied in the physical science. In particular, the subjects studied by *dynamics* in physics describe the world as *causally changing and dependently arisen,* which is nothing but the coming and going in the macroscopic world outside ourselves.

What is known as *nirvāṇa* is the transcendental state of viewing the world of continuous coming and going as it is, or it is concerned with what is realized by the ultimate path, where practitioners practice mindfully and transcendentally.

> *That which, taken as changing causally or dependently arisen,*
> *Is the conventional world.*
> *Taken as non causal and beyond dependence,*
> *That is said to be nirvāṇa*
>
> [Verse 9, Chap. XXV (Nirvāṇa)]

The two "That's" in the above verse correspond to the same phenomena. In other words, the very same world is *saṃsāra* or *nirvāṇa*.

There exist two attention systems in the human brain: the top-down attention and the bottom-up attention. In other words, our brain has dual (or multi) attention systems of neural network. It is interesting to realize that these enable multiple views of the same world.

The next question is whether *nirvāṇa* exists or not?

Buddha advised to abandon
Both existence and non-existence.
Therefore, it makes sense that
Nirvāṇa is neither existence, nor non-existence.

[Verse 10, Chap. XXV (Nirvāṇa)]

Those who long for liberation from real existence by means of *real existence* lack deep insight; those who long for liberation from real existence by means of *non-existence* also lack deep insight.

Both longing for something that exists and longing for something that does not are to be given up. However, *nirvāṇa* should not be renounced because *nirvāṇa* neither exists nor does not.

If there were something called "the existent," then one should distinguish *nirvāṇa* negatively from it and claim, "*Nirvāṇa is not the existent.*" If there were something called "the non-existent," then one should distinguish *nirvāṇa* negatively from it, and *nirvāṇa* would be definitely "not non-existent."

(22) *Nirvāṇa and saṃsāra: no difference*

The discussion here is framed again by the *tetra-lemma* (described in item (*20*)). It may be said that *nirvāṇa* is something about which a description could be given, or an alternative to the *saṃsāra* from which it is inherently different. If it were so, it would have to be either existent, non-existent, both, or neither. Taking into account the logical reasonings of several verses after the 10th, one concludes that none of the four possibilities for *nirvāṇa* can be asserted. Therefore, *nirvāṇa* is different from the objects that can be given any assertion. When we try to say (assert) something coherent about the nature of things from an ultimate standpoint, some unenlightened people end up talking nonsense [48]. Beyond all reasonings, with profound insight, we must observe those from the whole body.

Seeing the Tathāgata from the ultimate point of view, one can draw one of the most startling conclusions. We have already considered in verse 9 that the very same world is seen as *saṃsāra* (cyclic existence) and also as *nirvāṇa* (the ultimate). Precisely for this reason,

The saṃsāra does not have
the slightest difference from nirvāṇa.
nirvāṇa does not have
the slightest difference from the samsara.

[Verse 19, Chap. XXV (Nirvāṇa)]

Nirvāṇa is the *saṃsāra* as seen without attachment, without mental fabrication, or without delusion. The reason we cannot say anything about *nirvāṇa* as an independent entity is not that it is such an entity, but that it is *ineffable* and incomprehensible through language [48]. The natural world evolves on its own, without words, without written laws, serenely and solemnly.

It may sound strange to say that there is no difference between *saṃsāra* and *nirvāṇa*. But each is empty in inherent nature, so there can be no inherent difference. *Nirvāṇa* is, by definition, the cessation of delusion, grasping of desire, and reification of self, and the other (samsara) is that which occurs within the human brain. *Nirvāṇa* happens immediately after those brain activities are eradicated by practicing the noble path. That is why the Tathāgata encouraged people to follow the Path.

(23) *Nirvāṇa is the ultimate emancipation*
If there were anything at all called Truth (*Dharma*) that absolutely exists, there would be those who were the teachers of the Truth. However, no such thing exists. Thus,

Nirvāṇa is not something which can be captured,
That is cessation of mental fabrication.
No dharma was taught by Buddha
At any time, anywhere, to anyone.

[Verse 24, Chap. XXV (Nirvāṇa)]

The ultimate emancipation is nothing but the coming to rest of the root afflictions (desire, delusion, aversion, etc.) and the pacification of mental illusions. This itself is *nirvāṇa*. The coming to rest of mental fabrication resulting from abandonment of the basic afflictions, and hence total extirpation of the mode of thought, is the ultimate joy. The cessation of mental fabrications through not seizing objects of knowledge or knowledge itself is the ultimate emancipation [45, 46].

Scientifically speaking, even after the pacification of mental fabrication has been attained by *dhyāna* practice, the brain still maintains its activity and the body continues its physiological functions. Human activity never stops. Rather, the brain is immersed into its fundamental activity, which has been formed since the beginningless ancient times, and the whole body is maintained in a relaxed innate state by faint deep breathing. Eventually, in harmony with the surrounding environment, the whole body enters a serene state naturally. Then, personal existence ceases, and ultimate emancipation appears spontaneously. That is the state realized through the *dhyāna* practice. In *nirvāṇa*, everything evolves beyond language. According to Ācārya Nāgārjuna:

> *Nirvāṇa is neither be made extinct, nor be attained; neither be terminable, nor be permanent; neither ceases to be, nor comes to be.*

(24) *Nirvāṇa and worldly views*

Nirvāṇa is characterized by stillness, which transcends mental fabrications (eradicating mental notions). Metaphorically speaking, it is like the clean, tranquil stillness at midnight. This is expressed in English as "thusness" or "suchness" and in Chinese as 如是 (*rúshì*). Nirvāṇa is neither something that can be attained by extirpating (eradicating) desire nor something that can be attained through action, such as the fruit of moral striving, nor something that is permanent, such as what is not devoid of beings:

> *Nirvāṇa is said to be*
> *What can neither be made extinct, nor be attained;*

what neither be terminable, nor be permanent
what neither ceases to be, nor comes to be.

[Verse 3, Chap. XXV (Nirvāṇa)]

Worldly views: From the beginningless ancient times, desire, hatred, and delusion have been born of *nescience* and are caused by erroneous views. The root of all erroneous views is the belief that the self or the external world exist inherently. Such views bind us to saṃsāra. Once all afflictions such as desire, hatred and delusion have been eradicated, *nirvāṇa* emerges immediately. All things are ultimately *empty* in nature. Thus,

All phenomena are empty of inherent nature.
Therefore, what on earth such worldly views
*of permanence, self, etc. w*ould occur at all
by what reason, to whom, at what place?

[Verse 29, Chap. XXVII (Views)]

The self and all worldly phenomena are dependently arisen and hence empty in nature. There is no support for permanence and for self. If we bear in mind the emptiness of all phenomena, false worldly views would not occur at all.

(25) *Prostration to the Great Saint in gratitude*

I prostrate to Gautama Buddha
Who, through great compassion
expounded the Noble Dharma
enabling complete eradication of all views

[Verse 30, Chap. XXVII (Views)]

(26) *Additional comments on nirvāṇa as Ultimate Beatitude*

Nirvāṇa is abandonment of mental fabrication.
Nirvāṇa is ultimate beatitude, not something that can be grasped.

Fig. 9.2   Soaring King eagle.

Owing to this very nature, it is said that no dharma was taught by Lord Buddha. The function of perceptions as signs of naming (that is the mental fabrication) ceases. That is nirvāṇa. The ultimate *nirvāṇa* is nothing but the coming to rest of the root afflictions (desire, delusion, aversion, etc.) and the pacification of mental illusions. When verbal assertions cease, named things are in repose. The ceasing to function of discursive thought is ultimate beatitude. The coming to rest of named things by non-functioning of the basic afflictions (so that personal existence ceases) is ultimate beatitude.

Commenting on the paragraph of *the coming to rest of named things*, the translator (M. Sprung) of the commentary [45] wrote in the footnote (p. 262 of [45]): "This paragraph is the pithiest account of *nirvāṇa,* by the author Candrakīrti of the *Prasannapada*, turning on the notion as though the turmoil of the world in time were a distortion arising from human passions."

The coming to rest of named things as a result of abandoning the basic afflictions and hence totally eradicating the innate mode of thought is ultimate beatitude. Again, the coming to rest of named things by not seizing on objects of knowledge or knowledge itself is ultimate beatitude.

9.6 An Eagle Soars in the Sky. Is There Any Secret There?

The Tathāgata is like a king eagle soaring in the sky. When the Tathāgata is in *nirvāṇa*, the Tathāgata soars in the sky of emptiness (*nirvāṇa*), flying grandly with both wings spread out and soaring mindfully with one wing of Great Compassion and another of *insight*, or *wisdom*. The wind grasped by both wings is the joy of *nirvāṇa*, void of mental fabrication. Without staying at any fixed place, the Tathāgata Eagle does not perceive objects as signs of something existent, nor does it perceive any rigid "truth," either bondage or purification, that can be addressed to people.

The world as it is seen is something that is inexpressible in words and beyond languages. Behind the diverse complexions behind them, there should be an "intrinsic secret," the utmost secret that is the Tathāgata-secret. The following is an extract from the sutra "Tathāgata-Secret" (*Tathāgataguhya Sutra*) [45, 46]:

Shantamati!! During the night when the Tathāgata had awakened completely, achieving perfect awakening (*samyak-saṃbodhi*) with unsurpassed perfect enlightenment, and during the night when he passed totally into *Nirvāṇa*, and between (the two nights), Tathāgata did not utter at all, nor did he say any one syllable. Neither did he address anyone, nor does he address now, nor will he address later. Yet (however) all sentient beings perceive the voice of the Tathāgata variously according to their propensities (natural tendencies to favor something) and their beliefs as it issues forth in the conventional societies. They take it variously: "The revered one is teaching this doctrine for our benefit," or, "We are hearing the doctrine of the enlightened one." But, in truth, the Tathāgata has done nothing of mental conceptions (conceiving) and nothing of grasping existences, because the Tathāgata is freed from all thought of sorting out, from all mental fabrications and everything named." Furthermore, "Everything is inexpressible with words and beyond language, and is devoid of being, already in *Nirvāṇa*, and innately pure. One who knows them so is called a Bodhisattva, a Buddha.

In fact the Tathāgata had no secret. Everything is devoid of being. In nirvāṇa, everything evolves beyond language, and no word is appropriate to express something devoid of inherent existence.

Seeing an eagle soaring high up in the sky, ask whether the bird has any secret means of flying. The bird may reply:

"I am just soaring in the sky. I have no secret. I am soaring and gliding over the wind stream omnipresent in the sky. My flight is beyond language."

(Needless to say, birds are soaring for predation. This section is concerned with the dynamical aspect of soaring flights.)

However, the sentient beings living on the ground at the bottom of the sky have their own views about the bird's flight according to their propensities: "The bird is teaching how to fly in the sky for our benefit," or, "We hear the doctrine of the master of flight."

The Tathāgata, who is awakened with unsurpassed perfect enlightenment, is soaring in the sky of emptiness, *nirvāṇa*, flying grandly with both wings spread out and soaring mindfully with one wing of Great Compassion and another wing of insight, or wisdom.

9.7 Afterword: Concept of Transformation in NMK and Science

In the natural world, continuously changing phenomena are observed commonly and innumerably. Modern sciences intensively investigate those natural phenomena. For example, plant growth is studied by biology, and the evolutionary life of a star, from birth to later activity, including nuclear fusion within it and supernova explosion, is studied by astrophysics. The plant growth stages are as follows: from *seed* to *sprout*, then *vegetative*, *budding*, *flowering*, and *ripening*. The continuous sequence of states of a plant or a star is called *evolution*, or *transformation*. Scientifically speaking, the sequence of changing states is viewed as "the state transforms itself according to biological laws or physical laws," rather than "each state at an instant perishes before the next arises."

It is amazing to know that such a concept of transformations for human states was already studied in ancient India. The Sanskrit *pariṇāma* describes the transformation not only of the human mind but also of the human body and the process of sensing external signals. In Section 6.2, part (2), investigates how the *pariṇāma* within one's body results in direct realization of mental awareness. In fact, it is essential to realize the effect of the *life stream* occurring within one's body so that the practitioner acquires insightful awareness through *pariṇāma* (transformation). In

Section 8.7, too, *pariṇāma* was emphasized as an important mechanism of the yoga practice.

The evolutionary sequence is an essence of natural phenomena. However, if such natural phenomena are to be described with conceptional words, they must be inevitably expressed by negative expressions, as mentioned definitely by the Eight Negations in Section 9.3(2) ("Opening theme"), because those conceptional words are empty in essence, having no intrinsic nature, and are called "emptiness" or "things without self-existence," etc. This was a main point of the *śūnyatā* (emptiness) philosophy. That is exactly why it is mentioned in the opening of the masterpiece *NMK*.

All phenomena are arising together in a mutually interdependent web of interactions. Everything is co-arising dependently. Nothing exists independently. Nāgārjuna's *śūnyatā* theory began with the following verse:

Whatever is born with dependent origination

neither arises, nor ceases,	不生　不滅
neither permanent, nor terminable,	不常　不断
neither coming, nor going,	不来　不出
neither identical, nor distinctive.	不一　不異

The Buddhist view on our world is described by the *pratītyasamutpāda* (dependent origination). True way of things in such a world is characterized by the above negations with eight terms. Nāgārjuna's intention is to establish the absence of any substantial entity (*svabhāva*: own non-existence; natural state) underlying everyday phenomena. The term *śūnyatā* is a corollary of *pratītyasamutpāda* (dependent origination).

Let us next consider its significance with examples from *nature*. As mentioned above, the plant growth stages are: from the *seed* stage *to sprout*, followed by the *vegetative*, *budding*, *flowering*, and *ripening* stages. Consider trying to describe the plant growth using conceptional words such as *seed, sprout, budding*, etc. Although one can simply record the observed facts of the growth sequence or just show those with graphic drawings, we encounter difficulties when attempting to describe the growth sequence by their causal relationship — that is, to describe what are the cause and effect of plant growth. What happens in the natural

Fig. 9.3   Bean germination (from Google: "Germination").

world is free from conceptional wording. In fact, plant growth is a typical case of *dependent origination*. Dependent origination is characterized by *emptiness*: "Whatever arises in dependence on something else does not arise intrinsically." Plant growth is a sequence of emptinesses. In order to characterize those transformation sequences free of *conceptual fabrication*, only the Eight Negations will do.

Seeing the growth process of the bean seed at each stage of germination (Fig. 9.3), it is as follows:

> *The sprouting bean, neither has perished to nothing,*
> *nor has arisen (from others)*
> *The bean, is neither terminating, nor permanent;*
> *Neither has come from somewhere, nor has gone;*
> *Is neither self-identical, nor different.*

This describes what happens during the germination of a bean.

Next, let us consider another example of *Natural* phenomenon of much larger scales: the genesis sequence of a tropical cyclone. A tropical cyclone is originated in otherwise quiet tropical area by wavy disturbances over sufficiently warm sea surface which supplies abundant moisture. The moisture supply must be kept continuously for a period longer than 24 hours at least. Hence the cyclone is influenced by the earth rotation, causing its structure to become spiral (see Section 11.2). The cyclones emerge from the geo-atmosphere, and experience subsequently continuous transformation during growing periods, and finally decay into the geo-atmosphere by diffusing into the atmosphere of lower temperatures at mid-latitude regions. Thus, the lifetime of a cyclone is said to be continuous *dynamical transformations of emptiness*. There is nothing existing permanently. It is seen that the sunyata philosophy can describe natural phenomena adequately.

Overall the aim of sunyata philosophy is to highlight the phenomena that arise through rectifying philosophical concepts and theories. Since the foundation of the *śūnyatā* philosophy by *Acarya Nagarjuna* was laid about two thousand years ago, his philosophy was taken up seriously by many eminent philosophers in every country and every generation not only in India, but also in China, Tibet, Japan, Korea and Vietnam. In modern ages, Nagarjun's views have found strong resonance with Western philosophers such as Karl Jaspers and Ludwig Wittgenstein in the early 20th century, and has been studied by scholars such as Louis de la Vallee Poussin, J.W. de Jong and many others. The depth of his thought and the austere beauty of his philosophical poetry would justify that attention.

The *śūnyatā* philosophy is not an incoherent mysticism, but it is the *Mind* triumph beyond the very edges of language and metaphysics.

Chapter 10

Transmission of Śākyamuni Dharma by Bodhidharma

Fig. 10.1 Bodhidharma on a single reed crossing the river Yang-zi-jiang. This is a rubbing of the stone stele discovered at the Stupa (Fig. 10.7), Bear-Ear Mountain.

Gautama Siddhārtha looked up and saw the morning star when sitting for meditation in the predawn hours of the morning. It was precisely the sight that occasioned his enlightenment. With the sun crawling up from the eastern mountains, he called out into the dawn, crying, "I am awakened together with the whole of the earth and all its beings." This spontaneous outcry on seeing the morning star was the first and perhaps most penetrating revelation of the deep interrelation between a being and the surrounding world. The Buddha's enlightenment was itself a realization that a person, a star, and the entire earth are all manifestations of one being. A person is unable to discern its boundaries, and its existence is as vast as the universe and as long as forever. The eye with which the person looked upon the morning star was the one great organ of sight that has awakened all our dawns. [*Ineffable Buddha Mind; The Dharma Lamp*]

(*cf.* https://tricycle.org/trikedaily/sighting-morning-star/)

10.1   Sramana Mahākāśyapa and First Nun Mahapajapati

(a) *Wordless communication: first transmission of the Dharma Lamp*

Venerable Mahākāśyapa was one of the most revered of the Buddha's disciples and was foremost in *śramaṇa* practices, said to be "*dhuta practice* number one (頭陀第一)." Here, the Sanskrit word *Sramana* means a "seeker," and *dhuta* means "removed" or "shaken off." The *dhuta* practice signifies pursuing a scrupulous way of life as a monk or one who shakes off obstacles to advance on the noble path. Mahākāśyapa assumed a leadership position among the Buddhist sangha after Shakyamuni attained in nirvāṇa.

There is an impressive episode that shows a mind-to-mind communication between Buddha and Mahākāśyapa, which was a wordless story of the Flower Sermon (拈華微笑) (although it is not like a sermon, but captured intuitively). According to the Zen koan text *Gateless Gate* [67], the koan story of Case Sixth by Master Wu-men is as follows:

One day when Sakyamuni was in the Gridhakuta mountain (Vulture Peak), *Sakyamuni took a white flower with fingers and held it before the congregation. Everyone was silent. Only Mahakasyapa responded with breaking into smile.*

[*Ineffable Buddha Mind, The Dharma Lamp*]

This wordless Flower Sermon communicates the ineffable nature of *suchness* (*tathātā*: true nature). Mahākāśyapa's smile signifies the direct transmission of wisdom without words. Next is the *lion's roar* (see Section 10.2.2 (c)) of Śākyamuni:

"I possess the subtle mind of nirvāṇa, the wisdom eye of true dharma, which is the formless-form of reality and the subtle Dharma Gate that does not rest on words or letter, and mind transmission is beyond scriptures. This I entrust to Mahakasyapa."

Thus in the Buddhist tradition, Mahākāśyapa is regarded as the first person to have received the transmission of the Buddha mind or the mind seal of Buddha.

(b) *Life story of Mahākāśyapa*

Mahākāśyapa was born in a wealthy Brahmin family in Magadha, which during his time (6th century BCE) was a developed kingdom along the river Ganges in northeast India. He was named Pipphali, meaning "born under the pippala tree." Mahākāśyapa was a strict observer of the *dhuta* practice (thirteen *dhutaṅga*s [70]) denoted as follows: One (*i*) wears abandoned robes; (*ii*) possesses only three robes; (*iii*) eats only alms food; (*iv*) accepts alms food from everyone without skipping houses; (*v*) takes one meal a day; (*vi*) remains in the forest; (*vii*) remains or meditates beneath a tree; (*viii*) remains on the bare earth without shelter; (*ix*) sleeps at the allotted spot; (*x*) remains or meditates in burial places (remains among charnels); etc.

From his youth, he wished to practice a religious life [70]. One day after marriage, he noticed that when his farm fields were plowed, birds would come and pull worms out of the freshly turned earth. It occurred to him then that his wealth and comfort were purchased by the suffering and death of other living beings. His wife, Subhadra, also wished to practice a religious life. One day, she spread seeds upon the ground to dry and noticed that birds came to eat the insects attracted to the seeds. After this, the married couple mutually decided to leave the worldly life since both of them had no desire for the passionate life, and they became genuine ascetics. They parted initially to walk separate paths to find good teachers.

Two years later, Pipphali was told that Śākyamuni Buddha was Great Enlightened One who dwelled with his thousand disciples at Venuvana Vihara (Bamboo-Grove Monastery) in the suburbs of Rajagriha, the capital town of the Magadha Kingdom (see Epilogue, "Walking in the Valley of the Buddha"), so Pipphali followed the devotees to the Venuvana in order to listen to the Buddha. He was deeply moved by the virtues and wisdom of Buddha. One day, on his way home after Buddha's congregation, he saw the Buddha sitting under a tree as stately as a golden mountain. He prostrated himself before the Buddha and said, "Great teacher, please take me as your disciple." The Buddha said, "Pipphali, no one in this world is qualified to be your teacher. Do come with me as a fellow seeker."

(c) *First nuns*

Śākyamuni Buddha (563–483 BCE) was awakened to the Supreme Way at the age of thirty-five at Bodh Gaya (occasioned by the morning star), located south of the port Patna (now the city Patna along the Ganges), in 528 BCE (or 589 BCE according to Theravada schools).[1]

When his father, King Śuddhodana, passed away five years later, Queen Gotami (his foster mother) visited Buddha, who was staying at Nigrodha Park in their hometown of Kapilavatthu, and offered a number of robes. Then, she requested to be ordained as a nun. Buddha was silent for a long moment before he said, "It is impossible."

A few days later, the Buddha returned to Vesali. After his departure, Gotami gathered all the women who wished to be ordained. She told them:

---

[1] Some recent data (although not conclusive) proposed that Buddha's *parinirvana* is dated 20 years either side of 400 BCE.

"I know beyond doubt that in the Way of Awakening, all people are equal. Everyone has the capacity to be enlightened and liberated. The Buddha himself has said so. There is no reason he should not accept women. We are full persons too. We can attain enlightenment and be liberated."

"I suggest we shave our heads, put on the yellow robes of bhikkhus, and walk barefoot to Vesali where we will ask to be ordained. We will walk hundreds of miles and beg for our food."

Thus, the women agreed on a day to put their plan into action.

Early one morning, not far from the Buddha's hut in Vesali, Buddha's disciple Ananda met Gotami and fifty other women standing. Every woman had shaved her head and was wearing a yellow robe. Their feet were swollen and bloody. Hearing the report of Ananda, Buddha at last told Ananda, "Ananda, please go and tell Lady Mahapajapati (Gotami) that if they are willing to accept the Eight Special Rules, they may be ordained." Thus, all of them were accepted into the Order, and thus a seeker-group of nuns was formed.

This enabled Pipphali to call Subhadra, who had become an ascetic of another religion school. After all, Subhadra joined the Śākyamuni's Order of Bhikkhunis. Buddha gave Pipphali the name Mahākāśyapa. Mahākāśyapa means *great sage*. Kaśyapa was a common name in ancient India, referring to many different sages in the ancient Vedic and epic texts.

Mahākāśyapa lived as an austere monk even in his old age. Once Buddha held an assembly in the garden and asked Mahākāśyapa to sit with him. Mahākāśyapa prostrated himself before the Buddha and said, "Bhagavan (Buddha, Venerable One), I am not your chief disciple and hence I am not qualified to sit with you." The Buddha then described to the Order the boundless virtue of Mahākāśyapa, and added that, even without his help, Mahākāśyapa could still seek his own enlightenment and attain the stage of Pratyeka Arhat (Lone Buddha).

The Buddha advised Mahākāśyapa to stop the austere practice, but Mahākāśyapa said [70]:

"Bhagavan (Venerable One), I need to continue on in these practices as I am not as able as Sariputra, Maha Mollagana and Purna in Teaching. But I will not forget the kindness of the Buddha and can repay the kindness of the Buddha in this way. One who is propagating the dharma must set a good example to people, and virtue can be cultivated through the austere life. If one can get used to such an austere life, it shows one's ability for tolerance and the spirit of utter devotion to the dharma and the people. My practice of the austerities will indirectly help them. Bhagavan (Venerable One), for the consolidation of the Buddha's Order and for the sake of all living beings, I feel I cannot give up the practice. Please forgive my obstinacy."

After hearing that, the Buddha was pleased. He said to the bhikkhus:

"What Mahākāśyapa has said is correct. For the dharma, we must consolidate the Order. To consolidate the Order, we must allow some people to follow their practices if they wish to do those practices. People like Mahākāśyapa can inspire one in the practice of the dharma. Mahākāśyapa you may do as you wish."

According to Pāli texts, the Buddha often praised Mahākāśyapa for his ability as a leading person for the dharma.

Śākyamuni Buddha entered parinirvana at Kushinagar, located northwest of the port of Pataliputra (now the city Patna) in 483 BCE (or 544 BCE according to Theravada schools).

Ninety days later, a Great Assembly was held to agree upon what Buddha said. The First Council was formed by the leadership of Mahākāśyapa, Ananda, Anuruddha, Upali, and Purna. It was presided over by Mahākāśyapa. When the Buddha was alive, Sariputra and Maha Mollagana were the Buddha's right-hand men. Mahākāśyapa seldom participated in their activities, but he practiced the Way by himself diligently. After the two chief disciples passed away, and thereafter the Buddha entered nirvāṇa, Mahākāśyapa was the one who took the responsibility of leading the Order. Thus, it can be seen that his attainment and virtue were indeed great [70].

About thirty years after the assembly of the First Council, Mahākāśyapa attained nirvāṇa. He entrusted his duties to Ananda. Thus, Venerable

Ananda became the second recipient of the Dharma Lamp transmission. Thus, Venerable Mahākāśyapa is now regarded as the First Master of the Dharma Lamp transmission, succeeding the Dharma Lamp of Shakyamuni Buddha.

10.2 Dharma Lamp Lineage from Śākyamuni Buddha

Bodhidharma (c. 440–536) was an outstanding monk who played a unique role in the Dharma Lamp lineage and its transmission. First, we consider what impression Bodhidharma gave to Chinese people as the Twenty-Eighth Master in the Dharma Lamp (法燈) lineage. Next, we review how the Buddha Dharma was understood by a Chinese Buddhist in the early times (7th century CE) of Buddhism in China. The *Enlightenment Song* (證道歌) of *Yǒngjiā* Dashi (665–713) is an excellent text to see how Buddhism was accepted in China.

Bodhidharma was an Indian Buddhist monk who crossed the Bay of Bengal from South India to the Southeast Asian region and traveled as far as China (East land) in the first quarter of the 6th century. In those times, India was called the West-land in China. As described by his first disciple, "Master (Bodhidharama) brought Mahāyāna Buddhism to China, i.e. the Dharma Lamp of Sakamuni Buddha." He was the third prince of a great kingdom in South India. Having come to China, he stayed for some time in Guangzhou, in southern China, and later at the Shaolin Monastery (少林寺), in middle China. Bodhidharma is well known in connection with a story expressed by the compact phrase[2] "nine years facing a wall" (面壁九年).

Based on the philosophy and practice represented by this phrase, the Chinese Chan (禪 Zen) Buddhism originated in Shaolin and further developed in ensuing generations, and propagated over China and then toward Asian countries.

---

[2]The phrase "nine years" corresponds to his stay in Shaolin. Bodhidharma himself added three more years in China, so he devoted a total of twelve years (Chinese 一紀: orbital period of the planet Jupitar) for the benefit of people.

Bodhidharma is respected as the First Zu[3] (初祖 Founder Master) of Chinese Chan Buddhism. Chan (Zen) Buddhism was accepted as a new movement in China that brought about revolutionary change and enlightenment in the culture and lives of not only practitioners but also ordinary people in East Asia, including China, Japan, Korea, Vietnam, and other countries. In recent modern ages, Bodhidharma's Buddhism has gained strong resonance in Western people with the name "Zen."

The depth of philosophy and breadth of influence that Bodhidharma brought to the East land (Asia) is beyond one's imagination. The following signifies how deep the influence was. One of the most frequently asked questions (directed to practitioners for encouragement by Zen masters) is

What is the meaning of Bodhidharma's coming from the West-land (India)?
如何是祖師西来意 (*Ru-he shi zu-shi xi-lai yi*)?

This[4] is usually called a *koan* (公案 [68], an open question with a profound implication) proposed to practitioners in order to encourage them to directly seek the fundamental *suchness* (or *thusnesss*, the essence of Buddha's mind) and to ask about the essence of *suchnesss*. This is what Bodhidharma tried to transmit to the people of the East land. This approach resonated in the minds of the East land people, bringing about revolutionary change and enlightenment in the culture of the East land. Even now, after a thousand and five hundred years, the question

"Why did Bodhidharma come to the East land?"

is valid as a koan to us living in the present world of the 21st century.

---

[3] See Section 10.3.8, where "Zu (祖)" is defined in the form of verse, according to [54] in the section of the twenty-eighth Master "Bodhidharma," which was mentioned by Bodhidharma himself.

[4] This statement was first made by Master Hui-Ke (慧可), the Second Zu (二祖), who was the most distinguished disciple of Bodhidharma, and recorded in the introduction to the text "*Wash-Brain-Marrow* Sutra (洗髄経)" ([75], p. 11), which is cited as a footnote 5 in Section 10.3.3.

Bodhidharma was the key person who enabled the transmission of the Dharma Lamp of Śākyamuni Buddha from India (West) to China (East). Owing to him, the Dharma Lamp of Śākyamuni crossed not only the ocean but also the borders of nations. In addition, the transmission is still going on in the present world and will be passed to subsequent generations in the future.

10.2.1 *Bodhidharma as the twenty-eighth master*

In the tradition of Chan (Zen) Buddhism, they have an authentic lineage of transmission of the Dharma Lamp (法燈) from Śākyamuni Buddha. Buddha's principal disciple, Mahākāśyapa, was the first master of inheritance. In this lineage, the fourteenth is Ācārya Nāgārjuna, the fifteenth is Ācārya Kānadeva, the twenty-first is Ācārya Vasubandhu, the twenty-seventh is Ācārya Prajñātāra, and the twenty-eighth is Ācārya Bodhidharma. These are described in the Chinese Buddhist record 景德伝燈録 [54] concerning *Transmission of Dharma Lamp*, compiled in 1004 CE.

In the 7th century CE, Master Yǒngjiā Xuanjue (永嘉玄覚, 665–713) [69] mentioned in his masterpiece *Enlightenment Verse* (證道歌) in Chinese that Bodhidharma was the twenty-eighth master of Buddhism in the lineage of descent from Gautama Buddha. The author of the *Verse*, Master Yǒngjiā (永嘉), chants:

> *Mahakasyapa was the first, leading the line of transmission;*
> *Twenty-eight Masters flourished for transmission in the West (India);*
> *Dharma Lamp was then brought over the sea to this country (China);*
> *And Bodhidharma became the First Zu of Chan here.* [69]

In other words, it was acknowledged during the time (7th century) of Yǒngjiā (永嘉) that Bodhidharma was the twenty-eighth master in India and also the First *Zu* (祖) of Chan Buddhism in China.

10.2.2 *Enlightenment Verse* (證道歌) *of Yǒngjiā Xuanjue*

Master Yǒngjiā (永嘉大師) was a disciple of the Sixth Zu, Master Huineng (慧能大師: 638–713), in the lineage of the Bodhidharma school, but he was not an ordinary disciple. He is now known as the Great

Teacher, Dashi (大師), in Chinese. Why he is called the Great Teacher probably rests on his masterpiece *Enlightenment Verse*. It is an expression of enlightenment through his own verses that burst forth from his heart. It is certainly the *lion's roar*.

Two Zen masters, Huineng and Yǒngjiā, were Chinese Buddhists, and Huineng was also the thirty-third (=28+5) master, counted from the first, Mahākāśyapa, in India with respect to the Dharma Lamp transmission from Shakyamuni Buddha. Between the two masters, i.e. Master Mahākāśyapa and Master Yǒngjiā (author of the Verse), there exists a long time separation of about 1200 years. In addition, there exists a big geographical separation, one being in India and the other in China. In spite of these significant differences, it is remarkable that one finds little difference between their minds in terms of enlightenment (Buddha mind), i.e. the Dharma Lamp.

Let us see what the *Enlightenment Verse* of Yǒngjiā is in China after 1200 years. His song consists of 266 verses, with each verse composed of seven Chinese characters. The whole song is divided into 65 parts, each of which consists of 4 (or 6) verses. The verse cited in Section 9.2.1 is extracted from verses 169 to 172.

(a) **The *Enlightenment Verse* starts with a lion's roar**
The beginning verses of the Verse are as follows:

> *Have you not seen?*
> *A free seeker who has renounced "leaning from outside" and also "doing any fabricating action," and the one who does not dare to banish any delusion nor to seek any truth.*
> *Real nature of "Avidya" is nothing but the Buddha nature.*
> *Illusory "Empty-body" is nothing but the Dharma-body.*

Regarding the second line concerning the phrase "leaning from outside," *The Gateless Gate* [67] (in its preface) states: "Those which have entered through gates from outside are not precious treasures and what is gained dependently on causal conditions is always of the nature of arising and ceasing." Regarding the phrase *"fabricating action,"* Nāgārjuna's verse of item (9) in Section 9.5 states: "The karmic actions arise from mental fabrication. Fabrication comes to an end in the sphere of emptiness." The *"avidya"* in the fourth line means the fundamental *ignorance* (*nescience*) arising from the human physiological activity. In addition, Nāgārjuna's verse of item (11) in Section 9.5 states: "Not being dependent on anything other than itself, being already in nirvāṇa, nor being fabricated by mental fabrication, that is the real way things are." In other words, it is *emptiness* and is nothing but the dharma-body.

(b) Appearances of a Chinese *śramaṇa*

In his *Enlightenment Verse*, Yǒngjiā Dashi describes how Chinese Buddhist monks, namely disciples of Śākyamuni, are proud of themselves. See the following:

> An awakened Sakyan disciple says himself to be poor.
> The one is indeed poor in his body appearance, but not poor in their Way.
> As to the poverty, the body is wrapped with rags.
> As to the Way, a priceless jewel is stored in the mind.
> [Verse 12, *Enlightenment Verse*]

What do they look like in later generations? The following verses show that they are not so different from the *śramaṇa*s in ancient India.

> Going always in solitude, walking always alone,
> Those who have attained the ultimate path roam around the same Nirvāṇa Road
> Robe is antiquated, mind being pure. Bearing is naturally noble.
> Appearance is haggard, body being unshakable. People ignore him.
> [Verses 11, *Enlightenment Verse*]

The first two lines remind us of the discourse verses of the text *Sutta Nipāta* [72] (in Theravada Buddhism), all verses of which are thought to have originated during Buddha's lifetime. Let us cite two verses from section 1:3, "A Rhinoceros":

- A wise man walks alone like a rhinoceros,
  freely and independently,
  as a deer free from fetters in wild forests
  moves around wherever it wants for forage.
- A wise man walks alone like a rhinoceros,
  going everywhere in all directions, without any intention,
  content with whatever obtained, enduring all sufferings,
  without having fears.

In spite of the long time separation of about 1200 years since the times of Śākyamuni and Mahākāśyapa, it is observed that the essence of the Dharma Lamp is transmitted, although the details might have changed because of the large differences in cultures, climates, and nationalities.

(c) **Lion's Roar of enlightened ones**

The discourses of Śākyamuni Buddha resonate overwhelmingly among the congregations, as if they were the lion's roar. Yǒngjiā's Verse describes it as follows:

> Lion's roar, sounding fearlessly,
> Hundreds of wild beasts hearing it, their brains all are bursted
> A musth-elephant rushes about in confusion, losing its dignity.
> However, the heaven dragon, listening it quietly, is delighted.
>
> [Verse 18, Enlightenment Verse]

From the viewpoint of the *Sutta Nipāta* verses cited in (*b*), the last line may be written as "the forest *rhinoceros*, listening it quietly, is delighted." In the original Chinese expression, the musth-elephant is described as "scent-elephant (香象)." A musth-elephant is extremely dangerous to both humans and other elephants.

(*Musth* is a word of Persian origin and is translated as "condition of poisoning" in the languages of Northern India. The word *musth* is used to describe abnormal or drunken behavior, whether in humans or elephants. *Musth* can be defined as a periodic change in the behavior of elephant bulls, which can last from a few weeks to several months. This change is caused by testosterone hormones.)

(d) *Lifestyle of an enlightened one*

Where and how the enlightened one spends his life? The following verses illustrate that lifestyle:

> Deep in the mountains, dwelling among forest trees
> Secluded far from villages, there is stillness beneath a tall pine on the high peak,
> A rustic monk sits calmly and peacefully at the summit hermitage
> Retreat in profound quietude is highly serene.
>
> [Verses 87 to 90, *Enlightenment Verse*]

During the retreat, the monk is absorbed in concentration and in deep dhyana.

(e) *Song of enlightenment*

Dwelling in worldly life always requires attachment to goodness. It may plant the seeds of success, but will do no more than attaining the smaller fruits of human life. It is like shooting an arrow into the sky and then looking up. Losing its upward momentum, the arrow will fall back down. This observation means that future lives will not be as desired. Never worry about the branch-tips of things. Keep only the base-root of things firmly in place. Just one leap enables to land straight onto the awakened ground of the Tathāgata. Enlightenment is like the jewel-like moon floating in the deep blue sky of Lapis-Lazuli. It is like a wish-granting pearl. Thus,

> Having understood already this wish-granting pearl,
> Practices for the benefit of others and self are never-ending.
> River is shining brightly by the moon, wind is sighing in the pines
> Long night and clean atmosphere, what indeed it is!

Nirvāṇa! Isn't it serene? The practice of *dhyana* sitting is not only for the benefit of self but also for others because it is fearless for others. Motivated by great compassion, bodhisattva practices every effort with a spontaneous wish for the benefit of others, keeping in mind the bodhisattva vow "Never go to the other shore of Nirvāṇa unless all sentient beings have not been liberated."

(f) *Overnight guest*

Yŏngjiā Xuanjue (永嘉玄覚) is also known as "The Overnight Guest" (一宿覚) because he (玄覚 Xuan-*Jué*) stayed only one night at the temple of the Sixth Zen master, Huineng (慧能大師), although it was his first visit there. *Jué* is understood to denote the person "Xuan-Jue." It is interesting to remember that the Chinese *Jué* (覚) also means "awakening." But he already had a seed of awakening before visiting there. In fact, Yongjia was a learned monk of the Tiantai (天台) school (located near Shanghai 上海, Central China), which is different from the Zen school. Following the advice of his colleague master, Xuan-Jué (玄覚) had decided to visit Master Huineng (慧能大師) together with his friend, a long way to the south (Sháoguān 韶関, Guangdong 広東省), to see whether his understanding resonated with that of Huineng.

Having arrived at the temple, Xuan-Jué (玄覚) walked around the Master Huineng (慧能大師), who was sitting in his Master-Chair, three times by shaking his tin-stick monk-staff, and then stood staring at him. Huineng (慧能) commented on his lack of mindfulness:

"Usually, Buddhist *śramaṇas* (seekers) follow the three thousand manners, and have mindful behaviors of eighty thousand. *Virtuous Priest*! Where are you from?"

Xuan-Jué (玄覚) responded, "The incessant birth and death is a crucial issue. Time passes promptly and life is mutable."

Huineng: "Why you do not gain *birthless-ness* by your body, and why you do not understand the *motionless-ness* of Nirvāṇa."

Xuan-Jué: "Having gained bodily, it is nothing but *birthless-ness*. Having had insight into it, it is nothing but *motionless-ness*."

The Master said, "Good! Like that, exactly! Now, stay overnight here!"

Xuan-Jué accepted it. Then, prostrating with dignity, he asked the Master politely if he was allowed to leave the next day.

The Master said, "How quickly you do leave?"

Xuan-Jué then said, "Fundamentally speaking, there exists no *motion* at the other shore of nirvāṇa. How is there *quick-ness*?."

Master: "Excellent! Excellent!" Thus, Xuan-Jué is called the *Overnight Jué* (一宿覚).

[54] (Vol. 5).

Thus, any significant difference is not seen between Mahākāśyapa in 6 BCE, who was an ancient *sramaṇa* in India, and Xuan-Jué in 7 CE who was a *sramaṇa* in China.

10.3 Bodhisattva Bodhidharma

10.3.1 *Dialogue with Emperor Wu*

It is likely that Bodhidharma left India around 521 CE. After a long voyage of three years, crossing the Bay of Bengal and heading to the north for China, Bodhidharma finally arrived at the Chinese port city Guangzhou in south China in 524 CE. He stayed there for a couple of years and communicated with local people without being recognized as an outstanding Buddhist monk (except Indian residents helping him).

Perhaps it was in 526 CE or so that the local governor Xiao-ang recognized Bodhidharma as a man of noble origin from India. Emperor Wu (Wu-Di) of the Liang dynasty was informed about Bodhidharma's stay and Buddhist activities in Guangzhou. Wu-Di 武帝 (Di 帝 means emperor in Chinese) invited him to the capital Jian-kang (now Nanjing), which was at a distance of about 1000 km. It might have taken about a month or so for Bodhidharma to go to the capital. He arrived in Jian-kang in October. Sitting on a decorated vehicle driven by men, the Emperor came to greet him. It is said that there were interesting dialogues between Wu-Di and Bodhidharma during his stay in the capital. Now it is understood that those included the essence of Zen Buddhism. One day, the Emperor invited him to the court and hosted a ceremonial dinner for him.

Then, Emperor Wu-Di had the following dialogue with Bodhidharma, according to a well-accepted historical record [66] by Zen Buddhists:

"Venerable Bodhidharma, what sutras have you brought from the Westland India."

Bodhidharma replied:

"I have not brought a single letter of teaching." [57]

Next, Wu-Di said:

> *"I have constructed Buddhist temples, transcribed a number of sutras, and ordained many Buddhists as monks. What karmic merit is promised?"*

Bodhidharma: *"No merit."*
Wu-Di: *"Why do you say no merit?"*
Bodhidharma:

> *"These are the virtues fabricated in the mundane world. All that are fabricated in such ways will be in ruins in due time. There is nothing of intrinsic nature. Although it is considered to exist, it is just like a shadow which follows an object. Therefore, there is no merit."*

Wu-Di asked another question:

> *"What is the first principle of noble truth?"*

Bodhidharma :

> *"Being empty. Noble-less."*

Wu-Di asked: *"Who on earth is facing me?"*
Bodhidharma: *"No vijñāna."* (or *"I don't know."*)

[54, 66]

Emperor Wu-Di was not aware of what Bodhidharma meant. Actually, the Emperor was facing a distinguished monk who was, as it were, a bodhisattva, well aware of the Mahāyāna sutra *Lankavatara Sutra*, which contains the following passage:

> *From the night of Enlightenment till that of Nirvāṇa, Tathāgata*
> *has in the meantime not made any proclamation whatever.*
>
> ([52] Chap. LXI, p. 101)

Fig. 10.2 Bodhidharma crossed the Yang-zi-jiang River on a single reed. This is a rubbing of the stone stele discovered at the Bear-Ear Mountain.

Another sentence is given at a different place: "Tathāgata has in the meantime not uttered one word." This is consistent with the first reply: "I have not brought a single letter of teaching" [56]. As explained in detail in Section 10.3.2, the other subsequent replies are also consistent with the text description of *Lankavatara Sutra*. Later, Bodhidharma was given the title "Great Teacher" (大師 Dashi in Chinese).

The Dashi realized that there was no merit in staying at this palace. So, he left the kingdom of Liang, crossed the Yang-zi-jiang River (Long River), and eventually traveled to the north. A learned monk (Venerable Chih) informed Wu-Di that the Indian monk who just left was actually a bodhisattva, transmitting the Seal of the Buddha Mind.

10.3.2 *Bodhidharma transmitting the Buddha mind*

Bodhidharma's encounter with the Buddhist Emperor Wu was really a historical event. In fact, the essential part of the above dialogue was adopted as the first chapter of the *Blue Cliff Record* (碧巌録 [66]), a collection of 100 koans [68]. In the highly esteemed Zen text, the first chapter highlights the essence of the philosophy of Bodhidharma transmitting the Buddha Mind (Buddha Seal*)* [66]. An initial original version of the text was compiled by Xuedou Chongxian (980–1052), who collected 100 koan-episodes (*Hundred Verses on Old Cases*), taking from classic texts of Zen masters and historical documents. Later, a Chan master, Yuanwu Keqin (1063–1135), added his own elaborate annotations and commentaries on those Old Hundred Cases. Its first chapter, among the hundred, was titled "Emperor Wu Asks Bodhidharma," which described the essential part of the dialogue given in the previous section. Citing again the dialogues from there, Emperor Wu told Bodhidharma,

> "Who on earth is facing me?" Bodhidharma replied, "不識 Bu shi, no vijñāna."

The last word "不識" is usually translated as "*I do not know*" in English. But as evidently seen from the Chinese expression "不識', the original text does not contain the part "I". Hence, literal translation of the original "不識' becomes "*No Know*", or "*Does not know*", or more precisely "*No knowledge*", otherwise "*No consciousness*". This is because the first "不" denotes "*No*" and the second "識" denotes "*consciousness*". In order to understand the meaning behind, let us investigate the true meaning of "不識" from now.

This is associated with an essential point of this Koan-episode [68], which is now to be interpreted in detail in order to elucidate that their dialogues had a deep meaning. That is the reason why it was placed at the first chapter. The present writer presumes that the two authors of [66], *Xuedou* and *Yuanwu*, and the document writer *Daoyuan* who compiled the historical document [54], all were well versed in the Lankavatara sutra [52, 53, 65].

Emperor Wu later questioned Master Chih concerning the Indian monk. Chih said, "Your majesty. Do you know who this man was?" The Emperor said, "I don't know."

Tell! Is this ("I don't know") the same as what Bodhidharma said, or is it different? In appearance, it indeed seems the same, but in reality it isn't. People often are misled at it. When Master Chih questioned, Emperor Wu answered Chih sincerely and said, "I don't know." Seeing a timely chance for him, Master Chih said,

"This is the Bodhisattva Avalokitesvara transmitting the Buddha Mind Seal."

The Emperor felt regret, and he hoped that by sending an emissary, he could bring Bodhidharma back.

In the above translation, the Chinese word 識 has been given another word from the Sanskrit, *vijñāna* (consciousness). From the *Lankavatara Sutra*, of which *Bodhidharma* was well versed, *vijñāna* is explained as follows: "Every mind *vijñāna* [every consciousness] performs two functions simultaneously: i.e. perceiving and discriminating." One way to explain this double activity is to compare it to "surface waves" on the deep sea: with *vijñāna* as surface waves, *ālaya-vijñāna* is an analogous to the deep sea. According to the *Sutra*, the waves of changing and evolving consciousnesses are stirred on deep unconsciousness, *ālaya-vijñāna*, much like howling waves over the deep sea. Hence "No vijñāna" implies the nescience *ālaya-vijñāna*. *Ālaya-vijñāna* is interpenetrated with both ordinary consciousness and the deep unconsciousness (the fundamental truth, *emptiness*). That is the Ultimate Way (see Section 2.5(*vi*), "Manas and *ālaya-vijñāna*").

Thus, Bodhidharma's reply "不識" in the above dialogue is expressed as "No vijñāna," implying that he has no *discriminative perception*. This is Bodhidharma's response to the discriminative question of the Emperor. In addition, "不識" implies *ālaya-vijñāna*. *Ālaya* is conceived in the Lanka as being absolute. It is beyond the grasp of the individual, empirical consciousness. It is almost like emptiness itself, which lies behind all the vijñāna activities. The *Lankavatara Sutra* has the following statement: "If vijñāna is not born, there is no ignorance; if ignorance is absent, there is *no vijñāna*, and how can succession take place?" We should not forget that Bodhidharma was well aware of it.

Consider a language issue. From a grammatical point of view, "不識" is valid in Chinese, while "Not know" does not make sense in English. This reflects the difference in the linguistics of both societies. It is not a question of valid or invalid. The difference is concerned with a philosophical issue. The expression "I do not know" clearly describes that "I" do not know, not others, while in the expression "不識" there is no restriction concerning the subject that does not know. In order to reflect this aspect, the present writer has chosen the phrase "no vijñāna" to describe what Bodhidharma wished to tell. See that "no vijñāna" has the same structure as "不識." This is closely related to the philosophy of our world and the observation.

Another point is concerned with the philosophical background of Bodhidharma, who had wished to come to China, firmly holding his mission, purely to transmit the mind seal in order to arouse people and instruct lost people mired in mistaken views. One should not forget that he was already an elder monk, eighty or so. In addition, he was the twenty-eighth master in the lineage of the Dharma Lamp of Sakyamuni Buddha. In this regard, he was a unique person who could not be replaced by others. It is no wonder even if he thought that this visit was the last chance for him to transmit the Dharma Lamp to later generations. From this point of view, too, the translation "I do not know" of "不識" is not appropriate. He was already beyond the existence of a single person. Thus, the present writer suspects that the above dialogue using "不識" in two ways was a trap set by the authors of the *Blue Cliff Record* to guide lost people mired in mistaken views with great mercy. In fact, the part of dialogue including "不識" does not appear in records before the publication of [54] and [66]. In other words, the author of [66] added it intentionally.

To interpret the second point in more detail, consider the following fact. The dialogue was taken up in the very first chapter of the *Blue Cliff*

Record. In addition, in the Commentary section, the author Yuanwu was kind enough to make his own remark, as follows.

Xue Dou (the original compiler) wrote, *"How will all you monks capture the real point?"* Then, he added the following line.

Emperor Wu asked: *"What is the first principle of noble truth?"* Bodhidharma: *"Being empty. Noble-less."*

Emperor Wu: *"Who on earth is facing me?"* Bodhidharma: *"No vijñāna."*

This is where Xue Dou is excessively doting, doubling his efforts to help people. Now tell me, whether "Empty" and "No vijñāna" are the same or different? If you are someone who has personally understood completely, you will understand it without anything being said. Someone who has not understood completely will undoubtedly separate them into two.

In fact, in the previous chapter, in Section 9.5(21) ("Nirvāṇa and conventional world"), we saw a verse: "The samsara does not have the slightest difference from nirvāṇa." Owing to our *nescience* (i.e. *no vijñāna*) during the beginningless time, we suffer aimless wandering in samsara. On the other hand, in Section 9.5(3), we saw: "Dependent origination is characterized by emptiness" and "dependent origination is inherently of nirvāṇa nature." Hence, one may say that our *no vijñāna* does not have the slightest difference from *emptiness*. This is the Dharma Lamp of Sakyamuni Buddha, which Bodhidharma wished to transmit to the people.

Thus, it is now fully understood that the dialogue between Bodhidharma and Emperor cited in Section 10.3.1 reflects the overall perspective of the noble aim of Bodhidharma transmitting the Buddha Mind and reveals his noble mission of Dharma Lamp transmission beyond doubt. This is nothing but the objective of the author of the *Blue Cliff Record* [66], Xuedou (980–1052), assisted by the document writer Daoyuan, because the source of the dialogue might have been originally taken from the historical document [54], which was presented to the Emperor in 1004 and opened to the public in 1011, when the author Xuedou was 31 years old. It is very likely that both authors (of [54] and [66]) communicated with each other on the matter. The present writer suspects that the author Xuedou intended for the first chapter to expound the overall perspective of Bodhidharma's mission in the style of dialogues. In fact, the goal has been fully achieved.

All things are non-dual, as are nirvāṇa and samsara. For this reason, we should be aware that our existence is in the midst of realizing emptiness, no vijñāna, non-duality, and no-self nature. [cf. [52]: Chapter XXVII (pp. 66–68), *The Lankavatara Sutra: A Mahāyāna Text*, translated by D. T. Suzuki.]

> The *Lanka* does not forget to add that though the Buddha is known by so many different names, he is thereby neither fattened nor emaciated, as he is like the moon in water (surface), neither immersed nor merging. This simile is generally regarded as best describing the relation of unity and multiplicity, of one absolute reality and this world of names and forms [captured by *vijñāna*]. [52]

Bodhidharma was well aware of the above philosophy. This is understood from the fact that, fortunately, after a couple of years at Shaolin, he had succeeded in acquiring an outstanding disciple, Huike, to whom he could finally transmit the Dharma Lamp of Shakyamuni Buddha and hand him the text of the *Lankavatara Sutra*, which was the philosophy behind the dialogue.

10.3.3 *At Shaolin*

Having left the kingdom of Liang in the autumn of 526 and crossed the Yang-zi River (Fig. 10.2), Bodhidharma traveled north. After nearly a year, in 527, he arrived at Luoyang (洛陽) (Fig. 10.3), the capital of Wei (in northern China). According to a text [73], he visited the Yong-ning Temple (永寧寺) (a center of translation of sutras), located in its suburbs. Seeing the glorious monastery and the golden disks of the pagoda reflecting in the sun, he was amazed and very impressed, and said: "I have passed through numerous countries. There is virtually no country I have not visited.[5]

---

[5] According to the text [75], his disciple Hui-Ke (Second Zu) later wrote as follows:
我祖師大発慈悲、自西来東、餐風宿露、不知幾歷暑寒；航海登山、又不知幾歷
險阻、如此者、豈好劳耶?... 知有載道之器；可堪重大之托、此祖師西来之大義

Fig. 10.3   Bodhidharma traveling in a seeker-monk style. Bronze, Shaolin Temple 少林寺, China. Photo taken in 2009 by the author.

也。初至陝西、敦煌、遺留湯鉢于寺。次及中州少林、面壁趺跏九年。不是心息參悟、亦非存想坐成功。総因因縁未至。姑静坐久留、以待智人參求耳。... = Our Teacher had come from the West to the East. When eating, exposed to winds; when sleeping, covered in dewdrops. Never had recognized how many seasons passed. Sailing over the sea, walking in the mountains. Never had recognized how many storms attacked and difficulties came over. How does such a person get exhausted? ...

If one can find such a person who is receptive to the Supreme Way and walks on that Road, the person must be sufficient enough for entrusting the noble mission. This is the essential significance of the reason why the Great Teacher had come from the West (see Section 10.2). First he visited Shaanxi Province 陝西省 and Dūnhuáng 敦煌 (nearly 2000 km apart), leaving bowls (staying) at temples. Next, he went to Shaolin in the Central Plain (the middle reaches of the Yellow River, centered on the region of the capital, Luoyang), where he spent nine years sitting cross-legged while facing the wall. This was the practice of neither calming the mind nor achieving enlightenment through sitting meditation. However, because the opportunity of his noble aim was not yet met, he was sitting quietly for the time being and staying for long. Thus, he had been waiting for the wise man's visit to seek the Way ...

There is nothing comparable to the beauty of this monastery in our world. Even the Buddha realm (India) lacks this." He chanted "*Nham*" repeatedly and joined his hands together (合掌*Gassho*). Then, he stayed there for some days.

Later, he moved to the Shaolin Monastery (少林寺). The monastery was newly established in 496 CE, just before his visit, by Emperor Xiaowen of Northern Wei (北魏孝文帝), and is located on the slope at the foot of Sōng-Shān Mountain (嵩山) toward south-east of the capital city, Luoyang. The temple name refers to its location in the ancient grove (林 lín) of Mount Shaoshi (少室 山) in the hinterland of the Sōng-Shān Mountains, which have several peaks, including the Shaoshi and Taishi Mountains.

He had stayed there for nine years. There was a cave, now called Damo cave (達摩洞) on a hill. It is said that he spent the entire day sitting in meditation, facing a wall (Fig. 10.4). Seeing his practice (footnote 5),

Fig. 10.4.   Bodhidharma and Huike (Kyoto National Museum,1495), *Cutting Arm* by Sesshu (Japanese Zen monk) who visited China in 1468–9.

people called him Bi-Guan-Brahman (壁観婆羅門) [71]. Of scholars who had deep understanding of the dhyana meditation practice, there were none who did not have faith in him, but those who were obsessed with their own self-views came to blame him for the sitting-only practice [51].

In Luoyang, there was a prominent scholar named Shen-guang (神光 487–593 CE), about 40 years old, who was learned in all teachings of Con-zi, Lao-zi, and Zhang-zi. Hearing that a monk of high virtue (Great Teacher = Dashi) from India was staying at the Shaolin Monastery, Shen-guang came to Shaolin and paid homage to the Dashi, wishing to learn the deep philosophy of Shakyamuni. He expressed reverently to Dashi his wish to learn the supreme path.

Huike, Dao-yu, and other disciples practiced diligently, demonstrating superior aspiration and a desire to master the Supreme Way. However, whenever they tried to ask a question in the morning and evening, they received only silence. At that time, Dashi thought calmly whether they could sacrifice their lives for the dharma. All the disciples considered their encounter with the Dashi as great fortune for themselves. They served Dashi reverently, requesting to hear about the True Dharma.

One day, as snow was falling in December, Shen-guang stood up overnight in the garden in front of the place where Dashi was sitting and meditating (Fig. 10.5). The snow lay knee-deep. After midnight, Dashi opened his mouth:

"Why you have been standing in the snow?"

Shen-guang replied, almost in tears:

"Dashi, please open the Dharma Gate with great mercy and compassion, and provide relief for all sentient beings widely."

Shen-guang, taking a knife, cut off his *left* arm in front of the Teacher, who was standing in the snow. This could be a dramatic scene from a Japanese Kabuki play.[6]

---

[6] According to the text [62], Shen-guang did not have an arm already, because it had been cut off by a bandit (this could be a false story).

Fig. 10.5   ShenGuamg (later the 2nd Master Hui-ke), asking the Dharma in snow at shaolin Temple, by LiangJie; China, 13CE, Shang-hai Museum.

Later, a disciple of Huike wrote an article [56] explaining why his master Huike cut off the arm. He cited his master as follows:

"When I began practicing to attain the enlightenment here, I cut off one of my arms and stood in the snow from the first night (about eight p.m.) to the third night (midnight), and did not recognize that the snow came over my knee. In this way, I quested for the supreme path."

It is said that the snow around him was tinged with red at this time.[7]

Dashi (Great Teacher) was moved by the purity of his zeal and opened the Dharma Gate (the true path) for him to enter. [51, 67]

Huike asked:

"Dashi, would you please tell me kindly what is the mind seal of Buddhas?"

Dashi:

"The mind seal of Buddhas is not anything such as one can obtain from someone else."

---

[7] At the time the text [62] (645 CE) was published (a hundred years after the nirvana of Bodhidharma), the above story, i.e. cutting off the arm for the dharma, might not have been popularly known. The episodes of cutting off the arm and staying in the snow were written down in texts around 716 CE [56] and 713 CE [61].

Huike asked further:[8]

> *"My mind is not yet peaceful. Dashi. Have mercy to make my mind quiet and peaceful*
>
> (make me An-Xin 安心 in Chinese).

Dashi ordered: *"Show me your mind (Xin 心). I will make it quiet (An-Xin).*

Huike: *"I've searched for my mind, but am unable to capture it."*
Dashi delivered judgment: *"Quieting of mind is over for you."*

10.3.4 *Sacrificing oneself for the True Dharma (a Jataka story)*

It is likely that Shen-guang (神 [god] 光 [shining light]) decided to cut off his arm himself. He was determined to do it himself (Fig. 10.6). This

Fig. 10.6   Shen-guang (神光) cutting off his left arm [63].

---

[8] [67] Gateless Gate (1228 CE): No. 41. *Bodhidharma Pacifies the Mind.*

personal action should express a very strong internal wish for the dharma. It is a fundamental of Buddhism that the True Dharma cannot be given from one person to another, and that it is inherent in all sentient beings. In many Buddhist sutras, there are stories of sacrificing oneself for the sake of the dharma. In the present context, it is helpful to know the existence of such Buddhist texts. The following may explain what happened in reality.

Shen-guang was a great scholar who was well versed in all the teachings of Chinese classics and was studying Buddhist sutras such as the *Nirvāṇa Sutra* [64]. This Mahāyāna sutra had been newly translated into Chinese just before the time of Shen-guang. In this text, one can see the story of a young Brahman (the Historic Buddha in an earlier incarnation) ([64] (a)), as follows.

He was practicing at a place close to a cliff deep in the mountains when the god Indra, in the guise of a demon, approached him and recited a half-stanza of scripture:

> *"All things are impermanent;* 諸行無常
> *this is the dharma of continuous formation and decay (of this world)."*
> 是生滅法

Upon hearing only this much, the Buddha begged the demon to finish the verse, even promised to throw himself off the cliff in sacrifice if the demon completed the stanza. The demon finished the stanza, whereupon the Buddha inscribed it on a nearby tree,

> *Formation and decay have settled down.* 生滅滅已
> *Nirvāṇa is the supreme ease.* (*Liberation from formation and decay*)
> 寂滅爲樂

and then threw himself from the cliff. Indra quickly revealed his true form and caught the Buddha when he was going to fall down. The whole verse (whole stanza) is regarded as a compact expression of Buddhism and constitutes the basis of the *Nirvāṇa Sutra*.

Shen-guang (神光) wished to seek the Supreme Way like the young Brahman. He was a genuine seeker of the true dharma. He had the vow of bodhisattvas: namely bodhisattvas do not dwell in the world of ceaseless fabrications of forming and decaying by the fundamental wisdom, nor do they wish to dwell in nirvāṇa, in order to relieve sentient beings with great mercy (*mahākaruṇā*) ([64] and [65]).

Realizing the personality of Shen-guang (神光) and admitting his firm sincerity, Bodhidharma Dashi accepted him as a disciple and guided him with the best of his heart. Shen-guang was given the dharma name Huike (慧可 "wisdom will do"). In fact, Shen-guang was one of the most faithful Buddhist disciples (*number one and no second*). He was one of the wisest scholars and best Buddhists at the time. Without him, the Chinese Zen Buddhism would not have originated.

10.3.5 *How Bodhidharma guided his disciples*

The disciples Huike, Dao-yu, Tan-lin, and others served Dashi for some years, requesting reverently to hear about the True Dharma. Dashi (Great Teacher) was moved by the purity of their zeal and opened the Dharma Gate (the supreme path) for them to enter [51]. The Teacher guided them kindly to learn authentic Buddhism and helped them to study Buddhist literature. After four or five years, Bodhidharma Dashi handed to Huike the sutra *Lankavatara* [65] (four-fascicle version) and said,

> "*I have here the Laṅkāvatāra sutra which I now would pass to you. It contains the essential teaching of the Mind of Buddha, by means of which you should lead all sentient beings. They should be liberated naturally*" [56, 61]. He repeated several times, "*Use it for your future activity.*"

When the Teacher saw the disciples struggling to capture the points in the sutra, he helped them by explaining several times and advised them to use the sutra for their future [61].

10.3.6 *Bodhidharma was documented in China*

The text [51] includes reliable descriptions about Bodhidharma and his teachings that are considered to be authentic, thanks to Tan-lin who was well learned in Sanskrit. In addition, the text includes teachings of Huike (慧可), the most respected disciple of Bodhidharma. Although it has been partially known from old times in China, Japan, and Korea, this text [51] is a complete text, which was excavated in the beginning of the 20th century at the Dun-huang (敦煌) cave complex in the northwest of China. The text contains not only classic and reliable records but also the early philosophy of the school that has been saved from deformation by history.

Recently (2017), the present author (Kambe) was given (in fact, presented) by a Shaolin master (Ven. Shi Yan Lin) a book of a collection of historical publications in Chinese.[9] The book is titled 増演易筋洗髄内功図説 [75] and was edited in 1895 by Mr. Zhou of the Qing dynasty (清朝, 周述官). Having read it, the present author found that the book includes original texts made by Huike and other monks, in particular the Chinese translation of what Bodhidharma described.

Here, from the text [51], some important descriptions are extracted, which are very interesting.

(i) **Mind** (心, Xīn in Chinese) is formless, and so it is not existent. Mind is not existent of itself. But the mind functions ceaselessly, and so it is not nonexistent. Also, because the Xīn functions, yet there is no place of its existence, it is not existent. It is empty and yet it functions constantly. Therefore, it is not nonexistent. Because it lacks a self, it is not existent. It arises due to conditions, and so it is not nonexistent. This reality of mind (Xīn , Dharma) is termed as *suchness* (or *thusness*, or Ru-Shi 如是 in Chinese). Suchness (of mind) is indeed what Buddha expounded from time to time.

From the view of the present writer (Kambe), the term "Xīn" (心, mind) defined here was proposed by Bodhidharma himself, and Xīn is nothing but the ālaya-vijñāna expounded in the *Lankavatara Sutra.*

(ii) The mind's not belonging to anything is liberation. When one does not understand what mind (Xīn) is, the person pursues *dharma*. When one does understand it, *dharma* pursues the person. What Bodhidharma

---

[9]This is a Chinese book for healthcare and Qigong (気功), which describes traditional Chinese mind/body exercise and meditation, which involves slow and precise body movements combined with controlled breathing and mental focusing to improve balance, flexibility, muscle strength, and overall health.

tried to make clear was the quieting of mind (An-Xīn) by Bi-Guan [60,71], i.e. *dhyana* sitting by a wall.

(iii) When one is sitting, *dharma* is sitting. Not that the self is sitting, and not that the self is not sitting. Do not seek *nirvāṇa*. Why? It is because dharma is *nirvāṇa*. How could you use nirvāṇa to attain nirvāṇa? Do not seek dharma when sitting. How could you seek dharma (Xīn) by dharma (Xīn)! This is characterized as Dun-Wu (頓悟: immediate awakening, or all-at-once awakening) compared with Jian-Wu (漸悟: gradual awakening, or step-by-step awakening).

(iv) Xīn (mind) is like a tree or a stone, since it is tacitly silent, unaware, undiscerning, nonchalant about everything. Be like a fool (stupid). Why? It is because dharma lacks awareness and knowledge. The dharma of sitting gives us fearlessness *externally*. It gives us supreme peace *internally*. There is no difference between common man and sage in this respect.

10.3.7 *Transmission of Dharma and the robe*

Huike, as well the disciples Dao-yu, Tan-lin, and others, served Bodhidharma diligently for about six years. Bodhidharma now realized that the time had come. It was the year 532 or 533 CE. Bodhidharma:

> *"Originally, I came to this land to transmit the dharma. Now, I have gotten such disciples who mastered it with both mind and body. One has gotten my marrow of bone, another my bone, and the third my flesh. The one who gained the marrow is Huike, the one gained the bone is Dao-yu, and the one gained the flesh is Ni-zong-chi (Nun-Dharani)."* [74]

Then, Bodhidharma (Founder Zu 初祖) transmitted his robe to Huike (Second Zu 二祖) and told him to keep it as a seal of his faith in the dharma. Later, the robe was handed down from the Second Zu to the Third Zu, and so on till the Sixth Zu (Huineng).[10]

---

[10] During the time of the sixth Zu, HuiNeng (慧能, 638–713), it was the reign of the Tang Dynasty (唐朝). It is said that the *de facto* Empress Wu Zetian (武則天, 624–705) required HuiNeng to show her the robe.

It is said that, after having cut off his arm, Huike stayed at a hermitage located in Mt. Shaoshi, facing the Shaolin Monastery downward, in order to regain his strength and practice. Bodhidharma also visited him out of compassion and sympathy for the pain of Huike by the injury. It was the sixth year since he first met the Great Teacher (in 532 or so) when he received the Dharma Lamp from Bodhidharma Dashi (the twenty-eighth master). Since then, recuperating, he stayed at his hermitage quietly and faithfully like a recluse, keeping the Dharma Lamp lit.

10.3.8 *Last journey of BodhiDharma*

The following is a story of the last journey according to [76] published in 2006.

Bodhidharma Dashi transmitted Buddha Dharma to Huike and gave the sutra *Lankavatara* to him. Then, the Wei Emperor promoted Bodhidharma Dashi to the position of leader of the Buddhist school in the Northern Wei Dynasty. To perform his responsibility, Dashi set out on a pilgrimage of Buddhist patrol throughout the country, taking along his disciple, Monk Tan-lin (僧曇林)[11]. Unfortunately, this tour became the last journey of Bodhidharma, as stated now. It was 536 CE.

They started initially from the Shaolin Monastery located about 50 km southeast of the capital, Luoyang (洛陽), on the southern side of the Yellow River. On this day, after long pilgrimage, they visited the White Horse Temple (白馬寺) on their way. This was the first Buddhist temple in China, founded in 68 CE, when two Indian monks, Kasyapa Matanga and Dharmaratna, had arrived there on two white horses, bringing Buddhist texts, statues, and relics. Both Bodhidharma and Tan-lin payed homage to the temple as well as to the two Indian monks who had arrived here about five hundred years ago from as far as India.

From the point of view of the present writer, this was because Bodhidharma personally realized that he could not return to India anymore (because of his age). In addition, he himself wished to go to this Temple in order to say goodbye to India. He had come so far away from his home kingdom. His soul wished to fly back and return to India at this

---

[11] Tan-lin was a learned monk who in the past had engaged in the translation of sutras from Sanskrit.

time. The subsequent in India flow of events may imply the above, although constructed partly by imagination.

Having paid homage, both Bodhidharma and Tan-lin headed from the White Horse Temple to the Qian-Sheng (Thousand-Saint) Temple located within the Longmen Grottoes. The temple of their final destination was located at Longmen Grottoes (龍門石窟) along the Yi River. The straight line distance between the two places (Shaolin and Longmen Grottes) is about 20 km. However, they took a detour and walked indirectly to their destination.

Chinese monks stayed at this temple and worked on translating Buddhist sutras from Sanskrit together with Indian monks, including Bodhiruci, who was helping their translation. They welcomed the arrival of two great monks from Shaolin. Probably, Tan-lin also worked there in the past, assisting with translation of Indian texts.

Some days after their arrival, a local governor, Yang Xuan-Zhi, came to pay homage to Bodhidharma. The governor said to him: "Taking refuge in the Three Treasures: Buddha, dharma and sangha, this disciple has spent years. My wisdom still stays in the dark and gets lost in the Truth. May the Teacher be compassionate enough to open the Dharma Gate. The Teacher is the Zu succeeding the dharma in the West (India). What is the Way of Zu?" The Great Teacher responded with the following verse:

Even seeing evil, not trying to have hate;
Seeing goodness, but not trying to do anything;
Throwing away one's fool, but not trying to be wise;
Throwing away worldly living, but not trying to be enlightened;
Walking the great Way, Being without measure;
Regardless of fools and saints, Being transcendental.
This is said to be the Zu.

Hearing this, the governor pleaded to the Teacher: "May the Great Teacher show mercy to live in this world for a long time."

The Teacher confessed: "I am passing away, unable to stay long. Human personalities are different and diverse, making others often feel sad.[12] I am worrying."

---

[12] According to the texts [54, 57, 59], Bodhidharma received poisoned food packages [57] (poisoned bentos 著毒餉) from Bodhi-Ruci and Guang-tong Lushi (光統律師) of

The governor said to Bodhidharma: "I don't get it. Whom you mean? This disciple will fix it for you."

Bodhidharma said: "My mission is to transmit the Buddha Dharma and to benefit all sentient beings.

I will choose the path of passing away myself, rather than disclosing a story. It is not the dharma at all to choose the way of becoming safe by hurting him."

The present writer had known from before that there existed historical literature describing the above story in China. However, until recently, I did not believe it.[13] Now I believe it because of the understanding that is now given.[14]

The story given above and that to be given below show that Bodhidharma was really a bodhisattva, which I wish to emphasize here. Undoubtedly, there was a certain influence from the Jataka story of the *Nirvāṇa Sutra*, already described in Section 10.3.4. In fact, this sutra was newly translated into Chinese just before the historical meeting of Chinese scholar monks with Bodhidharma. The four-line verse chanted by the god Indra in the guise of a demon is nothing but the essence of Buddha's Teaching and also the *Nirvāṇa Sutra*. Bodhidharma's mission is to transmit the same essence, the Buddha Dharma, and benefit all sentient beings. Hurting anyone was out of his consideration. Let us continue the story of the last days of Bodhidharma.

When Bodhidharma and Tan-lin arrived at the temple, collaborative translation work was being carried out by Chinese monks and the Indian monk Bodhiruci. The sutra was *Lankavatara Sutra*. This had been already translated 100 years ago, in 443 CE, by the great monk Gunabhadra from

---

Northern Wei on several occasions. Bodhidharma, realizing what kind of food it was, took it, but it could not hurt him. Probably, the first package was safe, and the subsequent ones might have included weak poison. I guess, he took it, but probably regurgitated it. The next few times he was cautious. Even seeing this kind of evil action, he never tried to hate it. Probably he took it by selecting safer options. However, this time was different from the previous five times. Now that he had the best Dharma disciple and the Buddha Dharma had been already transmitted to him, it was likely that he thought to himself, "I will pass away and will not stay any longer," as was recorded in the literature above.

[13] In the tradition of the Zen schools in Japan, the *no-naming* of "who was the person" has been faithfully kept. I have never heard Zen teachers in Japan telling such a story. I did not believe it until recently.

[14] Therefore, it is reasonable to describe this story without naming who is the person because no one other than those involved in the case witnessed the event.

India, and is called *four-fascicle Lanka*. In fact, this was the text handed to Monk Huike by Teacher Bodhidharma. Hence, they knew the sutra very well. At this temple, the same sutra, but an expanded version called *Ten Fascicle Lanka*, had been translated, and it was being used as a text for monks.

On the next day of their arrival, Monk Bodhiruci came to see the two special guests from Shaolin, greeted them respectfully, and presented them with the *Ten Fascicle Lanka*. Then, the monk asked the guests to review the newly translated *Ten Fascicle Lanka*. The present writer supposes that Tan-lin knew the group of Chinese translators well because he had been one of translators in the past. It is likely that the two guests were implicitly invited by the group of translators. Then, Bodhidharma and Tan-lin spent ten days and nights to see the text and review the new publication carefully, including the accuracy of the translation and interpretation. As a result of their careful review, they were afraid. Both of them were full of misgivings concerning the influence of this new translation when it would be circulated and spread throughout Chinese society.

The Great Teacher Bodhidharma had received the Supreme Dharma, *Anuttara-Samyak-Sambodhi*, from the twenty-seventh master Prajnadhara in India, and was keeping Unsurpassed Perfect Enlightenment as the twenty-eighth master in Shakyamuni Buddha's lineage. What else could be done?

On a day when Tan-lin was on duty to teach internal arts of dhyana meditation and external body arts to all the monks of the Temple, Dashi stayed alone in his room. The following is just a conjecture made from what Tan-lin reported later.

During the day, Dashi was visited by Bodhiruci. Dashi must have told everything, without holding back, about Tan-lin's and his own misgivings about the new *Ten Fascicle Lanka*, including both the advice of rewriting it from the beginning to the end and the harmful effect of this new translation would have when it was circulated and spread throughout society. The response of Bodhiruci was beyond his imagination.

Tan-lin returned late at night and found Dashi panting heavily. Dashi was sitting with his legs crossed and his hands on the legs, bent forward. He seemed to be muttering:

"Don't be upset, please. After my nirvāṇa, don't make fuss about this."

Ealier, Dashi had said, "My mission is to communicate the Truth of the Buddha Dharma and to benefit all sentient beings. I will choose the path of passing away myself. I am unable to choose the way of becoming safe by hurting other persons. Rather I will hurt myself even if passing away."

Bodhidharma Dashi had now acquired the best dharma disciple, Huike, and the Buddha Dharma had been already transmitted to him. He had left India and come as far as China for this purpose. Dashi had thought that Monk Huike was such a supreme disciple that Dashi was more than satisfied to have him as successor of his dharma, and he would have no regrets even if his life was lost here. Considering the background of this story, I believe that this kind of action or non-action can be performed only by a bodhisattva, not by others. That is to say, I believe that Bodhidharma Dashi was really a bodhisattva.

Bodhidharma passed away in 536 CE[15] and was buried at Xiong'er shan (熊耳山 Bear-Ear Mountain, Fig. 10.7), west of Luoyang in He-nan Province. More than 200 years later, Emperor Dai-Zong (763–779) of the Tang Dynasty (唐皇帝代宗) recognized the greatness of the contribution of Bodhidharma and gave him the title of Great Teacher of Perfect Enlightenment (円覚大師) and the stupa, Pagoda of Emptiness (空観塔) [59]. Historical records describe that there was a formal funeral held by the Temple with the Emperor's condolences. It is said that a stupa monument was also constructed later at the foot of Bear-Ear Mountain.

The memorial pagoda was discovered in the late 19th century. Its photo (Fig. 10.7, taken by a Japanese Buddhist in 1999) shows Mt. Bear-Ear and the pagoda. A stone stele was also found there, showing that this is really the place where his body was placed. Later, the dharma of Bodhidharma is compactly expressed in the following four phrases:

No dependence upon words and letters, 不立文字
Special transmission outside the scriptures, 教外別伝
Directly pointing to the mind dharma, 直指人心
Having insight into suchness, Attaining Buddhahood. 見性成佛.

---

[15] Every year on October 5 [54], Zen temples in Japan hold a memorial service in memory of Bodhidharma Da-shi to commemorate the anniversary of his *nirvana*.

Fig. 10.7   (Left) *Photo of the Bear-Ear Mountain and Stupa of Bodhidharma* (taken in 1999). Now, at the same place, Kong Xiang Si (Temple of Empty Form 空相寺) was constructed in 2002 (see right). In the initial reconstruction phase, there was a significant contribution from the Japan Dharma Society, a group of Zen Buddhist monks in Japan. (Right) At the same Stupa in March 2009 (after 10 years from the Left), with the Master Shi Yan-Ci 釈延慈法師, Head of the *Temple* 空相寺. The author (on the left) visited the temple on his personal pilgrimage.

   According to some sources [57, 61], he himself claimed to be 150 years old. The present writer believes that his age was ninety or so. The reasoning is as follows. The references [54, 59] cite the verse (which was supposedly made by Bodhidharma, but this is not certain): 『江槎分玉浪、管炬関金鑠、五口相共行、九十無彼我』.This is a verse twisted by metaphors and analogy that only skilled persons can understand. However, it means, "Bodhiruci and Guang-tong (光統) prepared poison [see earlier footnote in this section]. They had a controversy with me about the practice style of Buddhists. 'Die' nullifies them and me." As read by persons skilled in Chinese, there exist two letters "九十" in the fourth part that are traditionally read to make "卆" in combination (a technical trick), meaning "death of a high-rank person." However, the simple-minded present writer reads "九十" directly without twist: the combination denotes just "nine ten," meaning *ninety*. From this, one can estimate that he was born in c. 440 CE or so.

   The present writer now realizes that the above four-line verse 不立文字　教外別伝　直指人心　見性成佛 includes the *key word* describing the styles of practice that caused the controversy. One is concerned with practice depending on scriptures, and the other with the practice conducted without words and letters. The difference was what had characterized the new practice style of Bodhidharma's Buddhism, which appealed to the Chinese Buddhists and brought about the rise of Chan Buddhism.

It is now nearly a thousand and five hundred years since the time of Bodhidharma. Ask yourself, which name is shining more brilliantly, Bodhidharma or Wu-Di (Emperor)? It is needless to say anything. However, it is said that, having heard the news of Bodhidharma's parinirvana, Wu-Di (who had missed the chance when he first met Bodhidharma) composed a memorial message expressing his deep sorrow and wholeheartedly praising the Dharma Wheel that Bodhidharma turned.

In fact, Wu-Di was one of the most respectful Buddhist emperors among Chinese emperors. He was the emperor of the Liang Dynasty for nearly fifty years, maintaining the vegetarian Buddhist policy in the early days of Chinese Buddhism in southern China. He was both lucky and unlucky, because he met a real bodhisattva without realizing that the one he met was a man of *suchness* (see (*i*) of Section 10.3.6). After Bodhidharma left the capital of Liang, Emperor Wu asked Master Chih about him. Master Chih replied, "This monk was the Bodhisattva Avalokitesvara transmitting the Buddha mind seal" (Section 10.3.2). Hence, Bodhidharma was already recognized as bodhisattva at this time, as also mentioned in [66]. In fact, this encounter formed the first chapter of the *Blue Cliff Record* (11th century) [66], a collection of a hundred koans [68] on Zen Buddhism. Thus, Wu-Di's name, like Bodhidharma's, is eternal.

10.4 Shaolin Monastery

10.4.1 *History*

The Shaolin Temple (少林寺 *Shao-lin si*) is a Chán Buddhist temple at the foot of the Song-shan (嵩山) in Henan Province of China. It was founded in 495 CE. It is a Mahāyāna Buddhist monastery that is also famous for its association with the Chinese *Shaolin Kung Fu* (少林功夫) arts. According to [62, 76], the Shaolin Monastery (Fig. 10.8) was built in 495

(a) Drawing (1671) [80]

(b) Recent photo (2017) taken by the author

Fig. 10.8　Shaolin monastery, viewed from a mountain.

on the north side of Shaoshi (the central peak of Mount Song, one of the sacred mountains of China) by the order of Emperor Xiaowen of the Northern Wei Dynasty. The *shào* (少) of "Shaolin" refers to Mt. Shaoshi, a mountain in the Song-shan mountain range, and *lín* (林) means "forest." With *sì* (寺) added, the name literally means "monastery/temple in the woods of Mt. Shaoshi."

When the capital of Northern Wei moved to Luoyang, not far from Mt. Song in 494 CE, a group of Indian monks too moved there. Among them was a monk of high virtue, Ba-tuo (Fa-tuo 跋陀 or Buddha 仏陀). Asked by the Emperor, he replied as follows. "Mount Song is a sacred place. It is now 400 years since the introduction of Buddhism in 68 CE, at the Yong-ping time of Han Dynasty (漢 永平), at the White Horse Temple just outside the city wall of the ancient Eastern Han capital, Luoyang. However, the temple has waned and Buddha images are covered with dust. Buddhism there is old fashioned." The Emperor responded immediately and ordered Ven. Ba-tuo to renew Buddhism and make a new Buddhist temple at the Song-shan [76]. Thus, the Shaolin Temple was founded.

Bodhidharma came to the Shaolin Monastery around 527, thirty years after its founding, and stayed there for some years since then (Section 10.3.3). Nowadays, walking along a foot path extending out from the temple and ascending upward the Wu Ru Feng (Five-Breast Peak), one arrives at a memorial gate (Fig. 10.9(b)) built in 1604 (during the Ming Dynasty). Behind the gate is Damo-dong (Dharma Cave), where Damo (Bodhidharma) sat (Fig. 10.9(a)) facing the wall for nine years, according to legend.

10.4.2 *Shaolin Kung Fu*

The text *Yì-Jīn-Jīng* (易筋経) [75], which had been left behind by Bodhidharma after his departure from the Shaolin Monastery and discovered within the walls of the temple after he had passed away, has the following passage:

> After Bodhidharma had faced the wall for nine years at Shaolin temple, a hole was made with his stare. He left behind an iron chest. When the monks at the monastery opened the chest, they found two books: the "Wash-Marrow Classic" (*Xǐ-suǐ-jīng* 洗髄経) and the "Change

(a)   (b)

Fig. 10.9 Dharma Cave (Damo-dong): (a) Damo sitting in the cave [77]. On the upper part of the gate, three Chinese characters are seen: "黙玄處 Mo Xuan Chu," meaning the Place of Deep Mediation, in the handwriting of a court calligrapher. (b) Gate and cave entrance.

Muscle-and-Tendon Classic" (Yî-jīn-jīng, 易筋経) within it. The first book was brought by Bodhidharma's disciple Huike, and disappeared; as for the second, the monks liked it. Practicing the skills therein, they fell into heterodox ways and lost its correct purpose of cultivating the Real. However, the Shaolin monks have made some name for themselves through their fighting skill; this is all due to their possession of this manuscript.

This may not be true, all but it is sufficient for giving us a certain hint.

Whether the Shaolin art skills can be attributed entirely to Bodhidharma is not clear. Although some Chinese records suggest that Chinese monks practiced martial arts prior to the establishment of the Shaolin Monastery in 497, the following also might be true. During his boyhood as a prince of a South Indian kingdom (Kshatriya), Bodhidharma was trained as a warrior in Indian martial arts. It is very likely that he was good at martial arts because he followed the traditional Kalaripayattu in South India.

Kalaripayattu (pronounced [kaɭərip: ajət: ɨ]) is an Indian martial art from the southern state of Kerala. It is practiced in Kerala and is one of the oldest fighting systems in existence. In January 2011, an International seminar was held at the Institute of Asian Studies (Chennai, India). On this occasion, the organizer had planned for demonstration exercises of two martial arts groups to be shown as an evening program. One group consisted of Kalaripayattu schools from Kerala and the other of karate schools (close to Kung fu) from Chennai. It was found that there was close similarity between them. A practicing member of the latter group said that 70% of their actions were common. The teacher of the latter karate school was trained in the USA under a Japanese teacher of karate from a traditional school of Okinawa, Japan. Karaté is a martial art developed in the Ryukyu Islands (Kingdom), which was the old name of Okinawa Islands in southern part of Japan before being absorbed into Japan. It is said that they had karate since old times. It is interesting to observe that there is a phonetic similarity between *karate* and *Kalari*. Obviously, there are similarities between karate and Shaolin martial arts, too. Although there is no direct evidence, there must be connections between the three martial arts: Shaolin, karate and Kalaripayattu.

In conclusion, the above story of three different martial arts lacks direct evidence of connections, although they are very similar. It is true that those are being played out in reality at different places in the contemporary world.

In the White Robe Hall of the Shaolin, there is an interesting mural painting depicting the Shaolin martial arts, called Shaolin Kung Fu (少林功夫) in Chinese. The wall is covered with frescoes portraying the daily life of the Shaolin monks, which are said to have been painted in the early 19th century. On one side is the description of the monks practicing their special ways of martial arts (Fig. 10.10). The two figures are those taken partly from the single big wall painting. Clearly, these depict martial arts being practiced peacefully between Chinese monks and (dark) Indians. The painting on the other side shows monks training themselves in combat patterns with weapons.

On the Kung Fu floor of the Guan-shi-yin Hall at Shaolin, there are a number of deep pits (Fig. 10.11). These are said to have been shaped by Shaolin monks during their daily footwork training.

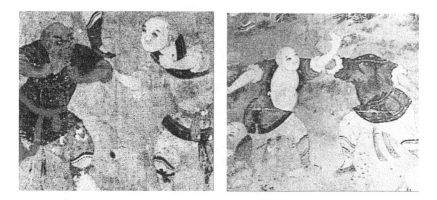

Fig. 10.10 Two parts from the mural painting of martial arts in "White-Robe Bodhhisattva Hall" at the Shaolin Temple.

(a)   (b)

Fig. 10.11 (a) Kung-Fu floor of Guan-shi-yin Hall at Shaolin; (b) its close-up. Photos taken by the author (2017).

10.5 Chan Buddhism (China)

In Section 10.3, we saw amazing and extraordinary stories of the initial phase in the transmission dramas of the Dharma Lamp from the Indian Bodhidharma to the Chinese Huike. Subsequent stories of succession from Huike to next generations are also noteworthy, resulting in the transition from a single lineage transmission to multi-lineages or parallel transmission. This transition enabled explosive development of the Buddhist school of Bodhidharma. Succeeding Chan Buddhism in China, Zen Buddhism emerged in Japan, as if transplanted from China.

10.5.1 *Introduction*

Bodhidharma succeeded the Dharma Lamp of the 27th Master, Prajnadhara, in the lineage descending from Shakyamuni Buddha in India, with Maha-Kassapa being the first successor of the Dharma Lamp [54]. In this way, Bodhidharma was the 28th Master in this lineage. The transmission from Indian lineage to Chinese lineage was recorded by a Chinese text, *Memoirs of Eminent Monks Continued* [62] (by Dao-xuan), as follows:

> "*Mahāyāna and Bi-Guan (Wall Contemplation) of Bodhidharma is the highest achievement.*"

This memoir was published in 645 CE, about a hundred years after the nirvāṇa of Bodhidharma (536 CE). Succession of the Dharma Lamp from Bodhidharma was hardly recognized at that time in the Chinese society. However, its succession was being carried out steadily by his disciples, *quietly, seriously, and intimately*, without words. This is the essence of the Dharma Lamp succession.

A text "Record of Lanka Masters and Disciples" [56] was discovered at the Dun-huang cave in the beginning of the 20th century. The original was published about 70 years (about 716 CE) after the above text [62]. Its author, Jing-jue (浄覚), wrote it as a private note in order to leave a record of what Bodhidharma told to future generations. The verse of faith in the opening paragraph expresses the circumstance of why the Dharma Lamp was transmitted steadily and secretly among the disciples, regardless of scant recognition of the contemporaries [56]:

> [A]: *Buddha nature is empty and no form. True nature (suchness) is quiet and no word*
> *Instructions by words of either voice or letters, those are all delusion-Zen.*
> *Nirvāṇa is a dharma of biting a flying arrow. It is a secret and is not taught to anyone.*
> *Mind penetrates, always functions silently. Dharma is open only to those of good fortune*
> *Persons of two vehicles do not know it. Non-Buddhists have never heard of it.*

Most of narrow views slander it. Wish to pray. Do not circulate it.
(See (*i*) of Section 10.3.6 for *suchness.*)

The author Jing-jue describes that the first master of his Chan school in China was Gunabhadra, since Gunabhadra had translated the *Lankavatara Sutra*, which was the four-fascicle *Lanka* that Bodhidharma handed down to Huike later. Gunabhadra (394–468) was born in Magadha, India. He traveled to China by sea via Sri Lanka and arrived at Guangzhou in 435 CE as an early-time Mahāyāna monk.

The Dharma Lamp of Bodhidharma is expressed compactly by "Wall contemplation and quieting mind" (壁観安心 Bi-Guan An-Xin), or more precisely by the following four holy verses (see Chapter 13):

[B]: *No dependence upon letters,* 不立文字
Mind transmission outside scriptures 教外別伝
Directly pointing to the living Mind 直指人心
Having insight into suchness, attaining Buddhahood. 見性成仏

This dharma was succeeded by Huike (慧可: 487–593), and then by Sengcan (僧璨). This is described in a stele (stone monument) [78] erected in 689, as follows:

Dharma teacher Bodhidharma brought from India the dharma of Shakyamuni to the east land, and came to the country Wei 魏 and transmitted it to Huike 慧可. Next, Huike transmitted it to Sengcan 僧璨, Seng-can transmitted it to Daoxin 道信, and Daoxin transmitted it to Hongren 弘忍.

[*can* in Chinese is pronounced as "ts-a-n"]
[*xin* in Chinese is pronounced as "sh-ee-n"]

Accordingly, Bodhidharma is called the First Zu (祖: "ancestor," "master," see Section 10.3.8) in the Chan Buddhism of China. Huike (慧可) is the Second Zu, Sengcan (僧璨: ? – 606) the Third Zu, Daoxin (道信: 580–651) the Fourth Zu, and Hongren (弘忍: 602–675) the Fifth Zu. In this way, the Dharma Lamp of Zen (Chan) Buddhism was succeeded. Next, it made its way to the Sixth Zu (六祖), Huineng

慧能 (638–713) (fifth Chinese Chan Master after Bodhidharma). The Dharma Lamp of Bodhidharma is expressed compactly by the following well-known verse [58]:

[C]: *I have come to this land with an ambition, that is*
Transmitting Dharma and relieving unsettled minds.
A single flower opens five petals.
Fruiting should be naturally achieved.

The first two lines describe the aim of Bodhidharma, who came to China. The last two lines are figurative expressions of flowering and fruiting. A person has a flower in his or her own heart, which opens its petals and comes to fruition. For people of succeeding generations, this verse overlaps the history described above. In fact, from the single flower of Bodhidharma, five petals of distinguished masters emerged as a succeeding lineage of the Dharma Lamp.

The story of transmission and development after the Fifth Zu (Hongren) was dramatic. There were about ten eminent Chan masters who succeeded the dharma of Hongren. During the Tang Dynasty, there was a lineage of Chan school that regarded Master Shen-Xiu (神秀: 606–706) as the Sixth Zu, and another lineage that regarded Fa-ru (法如: 638–689) as the sixth. Disciples of each lineage claimed that their own school was orthodox.

There was a difficult time initially, when the Dharma Lamp might have died out. This was followed by a period of rapid development. After that, each lineage competed to claim its school as orthodox. It is likely that verse [A] given above was written at this time. The sections to be given below describe this most interesting initial history of succession and development of the schools of Bodhidharma's Dharma Lamp.

10.5.2. *Huike* (慧可 *2nd) and Sengcan* (僧璨 *3rd)*

Bodhidharma Dashi passed away in 536 CE at the Qian-Sheng Temple (千聖寺), located along the Yi River (伊川) within the area of Longmen Grottoes (龍門石窟), south of the ancient capital Luoyang (洛陽). Not far from there, Huike (慧可) Dashi stayed on the bank of the Yellow River, living as if hidden. Needless to say, there were obstacles to his new style of Buddhist practice. However, former friends worshipped him for his fame, circulated an appeal of memory of praising, gathered monks, lay-men, and lay-women, and asked him for guidance. Huike responded with strong and deep talks and raised the spirit of the Dharma of the Great Teacher Bodhidharma [61, 62].

Later, owing to disorder within the government, the capital changed from Luoyang in Northern Wei to Ye-cheng (鄴都, 540) in Eastern Wei. Then, Huike Dashi also moved north-east to the new capital and expounded the mystical spirit of the dharma of his teacher, where the "mystical" is from the point of view of traditional Buddhists. He had no wish to open his own new school. However, there still existed such people in the capital who placed more importance on scriptures and letters than on the dhyana practice. Hearing his dharma talk, an acharya (leader) of a Buddhist school remarked, "This is a demon talk," because of its elusive-ness. Huike Dashi was still surrounded by an adverse atmosphere.

Huike Dashi (Great Teacher) cited a passage of the *Lankavatara Sutra*: "Those, who see the Buddha serene and behold Buddha beyond arising-or-ceasing, are without attachment or defilement in the world," and expounded that there had been no Buddha who had attained *awakening* without relying on zazen (dhyana sitting meditation) among Buddhas in the world of ten directions [56].

Someday, a layman of his age, about forty, came to see Huike Dashi. Presumably, it was 559 CE. Huike (慧可 487–593: Fig. 10.12) was already over seventy. From the references [54, 59], this is what happened. The layman bowed solemnly according to Buddhist formality and said,

"This disciple suffers paralysis. Dashi, please have mercy on the poor disciple to repel his sin."

Huike: *"Bring me your sin. I will repel it."*
Layman: *"I tried to get the sin, but cannot get it."*
Huike delivered judgment:

"Now, repelling of the sin is over for you. Please trust the three jewels: Buddha, dharma and sangha (bhikus and bhikushni), and take refuge in the three jewels."

Fig. 10.12   Second Zu, Huike 二祖慧可 (Shaolin).

Layman:

> *"Now, I see Dashi, who is a bhiksu, and hence understand who is a bhiksu. Please have mercy to tell me, what is Buddha and what is dharma."*

Huike:

> *"The mind is Buddha. The mind is dharma. Both treasures of Buddha and Dharma are not two separates. The third bhiksu treasure is the same."*

Layman:

> *"Now I realize for the first time that sin exists neither inside, outside, nor intermediate. Just as the mind is so, Buddha and Dharma are not two."*

Now, Huike was aware that the layman is a man of dharma talent. Then, he was shaved and ordained a monk. Dashi sensed that the layman is his treasure, and named him "Seng (僧)-Can 璨" [54, 59].

(*Can* 璨" in Chinese means brightness of a jewel stone. *Seng* (僧) means a monk. Sengcan is said to have been baldheaded and was called "Bare-Head Can" [59].)

In the years of 570s, there were two unfortunate anti-Buddhism movements (expulsion of Buddhism) by an emperor. This anti-movement might have been influenced by the disorder within the governing dynasty during the times of Bodhidharma Dashi's nirvāṇa. Both Huike and Sengcan (僧璨, 519?–606) had to bear the hardship of expulsion patiently. At the time of Buddhism's expulsion in 576, both of them escaped to the south and took refuge in a steep mountain, Si-kong-shan (司空山, Fig. 10.13(a)), in An-hui province (north of the Yang-zi River) and stayed, possibly, at the Second Zu's cave (二祖洞　Fig. 10.13(b)) and the Third Zu's cave (三祖洞　Fig. 10.14(b)) for several years, where the Second Zu is none other than Huike Dashi, and Sengcan was accepted as the Third Zu later (see below). Staying there, Sengcan recovered from his paralysis. Huike said to Sengcan (僧璨),

"Bodhidharma had come from India and transmitted True Dharma to me. I now transmit the dharma to you and give you this robe."

(a)   (b)

Fig. 9.13　(a) Si-kong-shan (司空山) in An-hui province (north of the Yang-zi River); (b) There is a cave of the second Zu (二祖洞) inside the house.

(a)   (b)

Fig. 10.14 (a) On the Rock, there are two calligraphies of Robe Transmission Stone (傳衣石), i.e. Observing Emptiness (観空); (b) Third Zu's cave (三祖洞) underneath the Rock.

Huike advised him to make it as a seal of faith in the dharma. It is believed that the transmission was carried out on the rock (Fig. 10.14(a)). Furthermore, he said,

> *"You must keep it and guard it carefully. Never let the Dharma Lamp be extinct."*

Huike then gave Sengcan the *Lankavatara Sutra* translated by Gunabhadra (Section 10.5.1) and told him to rely on the Sutra for practice. Presumably, Huike (with his left arm absent) was over 89 years old at the time (with Sengcan in his late fifties presumably). They were still suffering the crisis of Buddhism. This was in stark contrast to modern Zen Buddhism.

10.5.3 *Xìn Xīn Ming* (信心銘) *and Yuen-fu Temple* (元符寺) *of Huike Dashi*

Although there is no evidence, one can conjecture that it might have been just before the time of dharma transmission when Sengcan composed the verse *Xìn Xīn Ming* (信心銘, Verse of Faith-Mind), needless to say, under the guidance and instruction of Master Huike (慧可大師). It is an outstanding masterpiece and one of the most influential verses in Zen history.

This might be a verse instilling confidence in the belief that it is worth receiving the Dharma Lamp.

It consists of 146 lines of four characters. The first word, *Xìn* (信), of *Xìn Xīn Ming* nominally denotes *faith* or *trust*, but it essentially means *belief* in the *dhyana* practice *with the transmitted Robe for faith*, while the second word, Xīn (心), with a different accent from *Xìn* (信), denotes the mind (i.e. Dharma). From the view of the present writer (Kambe), the term *Xīn* (心, mind) here was proposed by Bodhidharma himself, and *Xīn* is nothing but the *ālaya-vijñāna* expounded in the *Lankavatara Sutra*. *Ālaya-vijñāna* denotes the unseparated ultimate state of both ordinary consciousness and the deep unconsciousness (the fundamental truth). See Section 10.3.2 and also Section 2.5 (*vi*). This is nothing but the Ultimate Way (Section 10.3.6(*i*)).

Xìn Xīn Ming begins with the following first four verses:

The Supreme Way is not difficult, 至道無難
Except for those holding preference/dis-preference 唯嫌揀擇
Only when freed from hate/love, 但莫憎愛
Penetrating, it is clear deeply. 洞然明白

The whole of *Xìn Xīn Ming* states that the practice of dhyana (信 with *faith-robe*) and the mind (心 of *ālaya*-dharma) are not *two*, i.e. they are *non-dual*. In fact, the last part of *Xìn Xīn Ming* states:

One is all (一即一切), All is one, without distinction (一切即一);
Dhyana and Mind are not dual (信心不二),   Non-dual are Dhyana and Mind (不二信心).

This is the true world, which transcends words and scriptures. In fact, Xìn (信, faith) and Xīn (心, mind) are not separate, and these non-separate two function together interdependently. On the basis of the Lanka, Sutra [52, 53], there was a controversy mentioned in 10.3.8. The verse *Xìn Xīn Ming* interprets the *key point* of their styles of practice which caused the controversy mentioned in Section 9.3.8.

The verse Xìn Xīn Ming had established the philosophical foundation of Chan (Zen) Buddhism, and determined essentially its future

direction. Its philosophy is characterized by EMPTINESS. It is said that Tathāgata had not made any dogmatic statement. Therefore, Shakyamuni's Buddhism is open to everyone, since no dogma is stated.

The following two points are highlighted.

Firstly, ordinary scholars studying scriptures hold the view that dhyana (practice) and mind (dharma) are two separate units. Those are called the "theorizer" in the *Lankavatara Sutra*. According to the sutra [52, in *Dharani*-s after Chap. 9]:

Those theorizer-s who are ignorant of the principle of non-duality are lost in the forest of consciousnesses. Seeking to establish their theories, they wander about here and there. The Mind with purity is the state of self-realization of the Dhyana. This is the Alaya-dharma (Tathāgata garbha) which does not belong to the realm of the theorizer-s. The Mind is primarily pure, but varieties of karmas and defilements accumulate owing to the manas and other consciousnesses, giving rise to the dualism.

Secondly, the Second Zu, Huike Dashi expounded that there had been no Buddha who had attained *awakening* without relying on dhyana meditation (zazen sitting). Later, Shen-hui 神会, a disciple of Hui-Neng 慧能, wrote:

Internally, the awakened Mind is the Dharma transmitted from a master to a disciple. Externally, the robe denotes the sign of Dharma Transmission. [58]

The composition of the verse *Xìn Xīn Ming* might have been motivated by Huike's wish that the Dharma Mind which Bodhidharma had transmitted to Huike should not be extinct. This must be kept with a desperate effort, as if their lives were hanging by a thread, as Hongren (Fifth Zu) told Huineng (Sixth Zu) [54].

After transmitting the Robe and dharma to Monk Sengcan, the Second Zu, Huike, said to his disciple:

"Please stay at deep mountains for a while. Do not go to cities for itineration. What I have done is just the transmission of the Buddha

Dharma received from the Great Teacher Bodhidharma. Do not suffer misfortunes of the world. Regarding myself, I have cumulative karmas in this world, and have to make atonement for those. I have to return to cities." [54]

What is meant by the atonement of Huike Dashi? The present writer guesses as follows. Firstly, Huike Dashi received the Buddha Dharma together with two volumes of holy texts (Section 10.4.2) from his teacher, Bodhidharma. Despite this, the Great Teacher suffered misfortune during his last days. This gave Huike great sorrow. In addition, concerning the practice styles of Buddhism, there was a fierce controversy with a certain group of Buddhists within the government (see Section 10.5.2), and Huie-ke Dashi believed (so the present author guesses) that his teacher sacrificed himself to protect the authentic style of Buddhist practice from the attack of the opponent group. Hence, Huie-ke Dashi resolved to preserve the authentic Buddhism for the people of the city. Proceeding just along the path of Buddha Dharma is the only way he can make atonement for the Great Teacher. Thus, he got off the mountain down to villages and towns, leaving his trusted disciple in the mountain to keep the authentic dharma safe. This view is consistent with what happened next, as described below now.

During the time of Sui (隋, 589–618), Huike Dashi, having survived the crisis of Buddhism expulsion, appeared again in the capital city Ye (鄴都, Handan, Hebei: 河北省邯鄲) around 590 CE (103 years old).

He was dressed shabbily. However, his mind was firm and noble. He was a man of resolve for the dharma. With his unusual appearance, he strolled down streets and markets. Making atonement means for him to proceed along the path of Buddha Dharma wherever he goes and to offer relief to save people from their sufferings [54, 57].

He went to any place he liked: *dive bars, slaughterhouses, streets of foul talk*, etc. He could be compared to a lotus flower in muddy water. As as a bodhisattva, he lived among people. One day, while he was engaged in domestic service in someone's house, a person in the town said to him:

"Do not be employed to labor by others, because you are a man of Dao (Way)."

He was regarded as a sacred person. He replied, "I am self-possessed. Don't worry" [54]. His actions were driven by his own internal mind springing out from the *self*, which is nothing but *śūnyatā*, i.e. Dharma. In fact, he was a man of *suchness* (see (i) of Section 10.3.6).

One day, it so happened that Huike Dashi was engaged in his dharma talk in front of a temple gate. Inside the temple (匡教寺 *Kuāng-Jiaò sì*), a senior monk (Dharmacarya Biàn-Hé 辯和) was also giving a lecture on the *Nirvāṇa Sutra* (涅槃経). Huike Dashi's talk attracted a large audience standing like forest trees, including monks and lay people who had attended the temple lecture. Venerable Biàn-Hé was upset with this. Huike Dashi had never imagined that this would bring him a deadly misfortune. In fact, the temple authority later made a false accusation to the local governor against Huike because of violation of public order. He was then arrested and overtaken by ill-fortune (punishment of beheading). However, Dashi received it calmly and in a self-possessed way [54]. At the time (593 CE), he was 106 years old. The future of the dharma lineage seemed to be hanging by a thread. Fortunately, this time, too, misfortune was avoided.

The misfortune of Huike Dashi was reported immediately to the Emperor of the Sui Dynasty. The Emperor mourned for the sake of Huike Dashi from the bottom of his heart, and his court expressed sympathy as a whole. This event helped to open a new door for the later development of Buddhism [57]. In the year 642 CE, 50 years after the ill-fortune of Huike Dashi, a temple (the Yuen-fu 元符寺, renamed in 1100 CE) was constructed at his last resting place in Hebei Province, 河北省磁州. Sacred relics of the second Great Zu Huike were enshrined in the pagoda constructed in 732 CE within the temple. The photograph of the pagoda in Fig. 10.24, taken in 1945, is of the structure re-erected in 1057 CE [79].

In 592 (or 593) when the Third Zu, Sengcan, was staying with his colleague monk at the Shan-gu-si temple (山谷寺 Mountain Valley Temple, Fig. 9.15(b)) in Wan-gong-shan (皖公山) in An-hui province, a novice named Daoxin (道信 about 14 years old, later the Fourth Zu) came to see Sengcan and bowed solemnly. He asked for the teachings of the Dharma Gate that lead to "liberation":

(a) (b)

Fig. 10.15 (a) On the Rock, the calligraphy reads "Liberating Bindstone" (解縛石). (b) Stupa at the Shan-gu-si (山谷寺 Mountain Valley Temple).

Sengcan Dashi asked, *"Who binds you?"*
Daoxin: *"Nobody binds me."*
Dashi: *"If nobody binds you, what sense does it make for you to seek liberation furthermore."*

There is a rock at the Shan-gu-si temple (山谷寺) with the calligraphy "Liberating Bind Stone" (解縛石) (Fig. 10.15(a)).

After exchanging this kind of dialogue, Daoxin (道信) followed Sengcan (僧璨) and practiced diligently according to his teaching for nine years. The Third Zu, Sengcan (三祖僧璨), now realized that the opportunity for dharma transmission had matured, and he gave the dharma and the Robe to Daoxin, and said:

"In the past, Huike Dashi (Second Zu) transmitted me the dharma and the Robe. Now I transmit the dharma and the Robe to you. I will now go to the south to stay at the mountain (Luo-fu-shan, 羅浮山). Now that there is no reason for me to stay here." [54]

Huike Dashi was a real bodhisattva. This is demonstrated by the fact that the modern world, after more than fourteen hundred years, still benefits from his wisdom unlimitedly.

10.5.4 *Daoxin (道信 4th) and Hongren (弘忍 5th)*

In 605, an imperial ordinance was issued by the governing Sui Dynasty (隋朝), which formally permitted renouncement for Buddhists. Daoxin (580–651) was appointed as the formal chief monk of a temple managed by the governor of Ji-zhou (吉州). It was the first time that a Chan Master of the Bodhidharma lineage was publicly recognized. This was a great event in the Bodhidharma School of Buddhism.

> The policy of Sui Dynasty in favor of Buddhism influenced Japan, reaching across the sea. At that time, Prince Shotoku (Shotoku-Taishi, 聖徳太子) was the Prince Regent for the court of Empress Suiko in Japan. The policy of pro-Buddhism of the Sui Dynasty encouraged him to construct the temple Horyu-Ji (法隆寺) in Japan. Since then, Buddhism in Japan flourished and expanded throughout the country.

One day, in those times, when Daoxin (Fourth Zu 道信) visited Huang-mei district (湖北省黄梅県), he met an odd boy (of about seven years old) and asked him a question.

> Dashi: *"What is your surname?"* Boy: *"My surname is not usual."*
> Dashi: *"What is it?"* Boy: *"Buddha nature."*
> Dashi: *"Don't you have a surname?"* Boy: *"Because my surname is empty."*

Now, Dashi realized that the boy is unusual, and sent his attendant to the house of his parents. By the permission of his parents, Dashi ordained the boy and gave him the name Hongren (弘忍 601–675) [54, 55].

In 624, Daoxin (Fourth Zu at the age of 45) moved to the temple Zheng-jue Si (正覚寺 Right-Einlightenment Temple), i.e. the Fourth Zu Temple (四祖寺) (Fig. 10.16(a)), located at the west of the Double-Peak Mountain (双峰山), north of the Yang-zi River, in Hu-bei Province (湖北省黄梅県, 蘄州). He stayed there for more than 20 years and attracted many students. Daoxin said to his disciples:

> *"Devote yourselves to sitting. Sitting is fundamental. Do practice for three years, or five years. Having minimal food to prevent starvation,*

just close gates and strive to 'just sit.' Do not read sutras. Do not have conversations. Before long, you will be a useful man. Just as a monkey eats fruit of a chestnut by winkling out, you could get it by sitting. Such a person is rare."

Furthermore, when he was asked from his disciples,

"Who is the person who receives the Dharma?"
Daoxin replied with sighs, *"Hongren would do"* [61].

Hongren (弘忍 5th) was a man of few words by his nature and often regarded as unimportant by his colleagues [61]. Usually, he wished to render labor service for others, and followed others while keeping a low profile. Daoxin Dashi (Fourth Zu) recognized him as a man of dharma talent. During the day, Hongren spent his time running errands of others. At night, he did zazen sitting till next morning. Thus, he spent his years without living in idleness. He did not read sutras or texts, but understood everything he heard. Hongren received kind and appropriate guidance from the Fourth Zu, Daoxin 四祖道信. There is a cave of dharma transmission inside the white structure seen in Fig. 10.16(b).

Daoxin Dashi passed away in 651 at the age of 72. After receiving the dharma and the Robe from the Fourth Zu, Daoxin, Hongren (弘忍, 602–675) stayed at Zhen-hui Si (真慧寺), i.e. the Fifth Zu Temple(五祖寺), at the East Mountain (東山, Fig. 10.17(a)), located to the east of the Fourth

(a)　　　　　　　　　　　　(b)

Fig. 10.16　(a) Gate of the Fourth Zu's Temple (四祖寺), Huan-mei, Hu-bei. (b) There is a cave of Dharma transmission inside the white structure.

(a)   (b)

Fig. 10.17   (a) The Fifth Zu Temple (五祖寺), Huan-mei, Hu-bei. (b) Stupa of the Fifth Zu.

Zu's mountain. He attracted many men of ability. In ten years or so, his sangha developed into a most active Chan group since the time of the founder, Bodhidharma.

Lay people far in the capital Luo-yang said,

"There are many awakened persons in the East Mountain (東山) in Huang-mei district (湖北省黄梅県)." [56]

One asked the Fifth Zu, Hongren 五祖弘忍:

"To learn the dharma, why do you keep staying in a mountain, rather than staying in a city or village."

Hongren replied:

"Good woods used for structures are obtained from deep mountains, not from places of human habitation. Out of the way, trees would not be cut away by ax or hatchet, and each tree grows into a big wood. Then afterward, those are used as useful ridges or beams. Thus we know that if our minds live in deep valleys by keeping ourselves from worldly affairs and if we maintain ourselves within mountains, our mind is stabilized naturally and becomes quiet."

In this way the Fifth Zu, Hong-en, kept quiet and continued clean sitting. In most cases, he taught his disciples without uttering a word [56]. The Fifth Zu passed away in 675.

10.5.5 *Huineng* (慧能 *6th*)

Huineng (638–713) was born in Xīn zhou, in (southern) Guang-Dong Province (広東省新州). The place was a newly developed area, which is known from the word Xīn, meaning "new." It is said that his father had been exiled from (northern) Hebei Province to the south. He lost his father when he was young.

After having grown up, he helped his aged mother in difficulty and sold firewood at the marketplace. One day, near a customer's house, he heard a Buddhist sutra being recited and noticed a phrase that was from the *Diamond Sutra*:

応無所住 而生其心 Yīng wú suo zhù, ér shēng qí xīn
Just not staying in attached, then the Mind is born spontaneously.
Our mind moves around, this way or that way. The mind is born spontaneously if it does not stick to anything and liberated from attachment.
The Mind is understood as in Section 10.3.6 (*i*).

Huineng's mind immediately resonated with the phrase (*Platform Sutra* [58]), and he asked the reciting monk about the sutra and the temple where he had learned it. Huineng decided to visit the Fifth Zu, Hongren (五祖弘忍), at the East Mountain of Huang-Mei, where the monk stayed. The customer gave him some silver to provide for his aged mother. After traveling for thirty days on foot, he arrived at Huang-Mei Mountain, where the Fifth Zu was presiding [81]. At last, he paid homage to the Fifth Zu, Master Hongren (五祖弘忍) [58].[16]
Hongren Dashi asked: "*Where are you from*? What are you seeking for?"

---

[16] It is said that, at the time, Hui-neng was 32 years old [54], or 24 years old [82].

Huineng answered: *"Being a farmer from Xīn zhou of Ling-Nan, I have come here to pay homage to the Master. I am seeking for the Buddhahood, not others."*

(Ling-Nan 嶺南 means the area south of the Nan-Ling-Shan mountain.)

Dashi: "You are from *Ling-Nan, hence a barbarian.* How you are able to attain Buddhahood?"
Huineng answered,

"There are *South and North* for people's living places. The Buddha nature has no *South and North. This barbarian is not the same as Your Reverend Master, but of the Buddha nature what difference is there?"*

Hongren Dashi realized deeply that Huineng is a man of great dharma talent. But he was still a layman, and appointed him to work as a servant for rice milling.

One day, the Fifth Zu, Hongren, announced to all the monks of his sangha (said to be more than seven hundred disciples):

"I have something to say to you. For people in the world, an essential matter is the one of birth and death. All day long, you have to try to get out of the bitter sea of birth and death. If you are confused about yourself, how can the mercy opportunity save you? "Each of you go back and look into your own true nature of Mind to compose a verse. Submit it to me so that I may look at it. "If someone has awakened to the essence of Mind, the robe and dharma will be passed on, and the one should be the Sixth Zu."

(A passage from the first chapter of [58].)

All the monks said to one another, *"Reverend Shen-Xiu* (神秀上座) *is our senior instructor and teaching transmitter. Certainly, he should be the one to obtain it."* Shen-Xiu then thought to himself, *"The others are not submitting verses because I am their teaching transmitter. I must compose a verse and submit it to the* High Master. *If I do not submit a verse, how will the Master know whether the views and understanding in my mind are deep or shallow?"*

Thus, only the Reverend Shen-Xiu (神秀上座) responded to compose a verse. To present his own view, he hesitatingly wrote it on the garden wall anonymously in the middle of the night. It was as follows.

The body is a Bodhi Tree, 身是菩提樹,
The Mind is like a bright-mirror stand. 心如明鏡台
Brush off diligently from time to time. 時時勤拂拭,
Let no dust be collected 勿使惹塵埃

Without being informed of this event, Huineng was working at the rice-milling house like a servant dressed poorly. After a couple of days, a novice-boy passed by the rice-milling house reciting this verse. Hearing the voice, Huineng immediately captured the verse and understood that it had not captured the truth of mind. Huineng (said to be illiterate) then asked a resident officer to write his own verse on the wall. It was as follows.

Bodhi has no tree intrinsically 菩提本無樹,
The bright-mirror has no stand either. 明鏡亦非台
There is not even a single existence inherently. 本來無一物
Where could any dust be attracted 何處惹塵埃

Seeing this verse of Huineng, all were shocked, and realized: "Amazing! One cannot judge a person by the looks or the working position alone." The above verse of Shen-Xiu is based on the view of inherent existence of the *body*, *mind*, *bodhi tree*, *mirror stand*, etc. However, Huineng's verse is based on the *śūnyatā* view, in which inherent existence is negated and no attachment to things is addressed. This is consistent with the phrase of the *Diamond Sutra*: "*Just not staying in attached, then the Mind is born spontaneously* (応無所住 而生其心)."

Seeing Huineng's verse and the response of his sangha members, the Master himself was terrified that there would be a reaction if he accepted Huineng's verse. Given the current order of the sangha and the fact that Huineng was still a lay person, it was likely that Huineng would be hurt by those who refused to accept him as the Sixth Zu. (Those people might be supporters of Reverend Shen-Xiu, but Shen-Xiu himself might have no control over them because he was just a scholar monk, not a politician.)

Next day, the Fifth Zu, Hongren, came to the rice-milling house of Huineng and saw Huineng, who was milling rice by stepping on hulled rice with a weight-stone tied around his waist. He said, *"A seeker of the Way would forget his self-body for the dharma. Is this not the case?"* [58].

The same night, the Master invited Huineng to his room and expounded the *Diamond Sutra* to him. When the Master came to the point of 应無所住 而生其心, Huineng awoke immediately:

> *How unexpected? The self-nature is originally pure in itself.*
> *How unexpected! The self-nature is originally complete in itself.*
> *How unexpected! The self-nature can produce the ten thousand dharmas.*

Thus, Huineng 慧能 was accepted as the Sixth Zu by the Fifth Zu, Hongren, and the Fifth Zu transmitted the Robe and the Bowl to Huineng. It was secret because there were a number of disciple monks who did not accept Hu-neng. They were unaware of the essential difference between the two verses. Master Hongren suspected that there might be adverse reaction among monks. The same night, Huineng had to escape to Ling-Nan, to the south, by crossing the Long River [58, 82].

In fact, there were hundreds of pursuers to take back the Robe and the Bowl from Huineng, who was carrying them. On his way to the south, Huineng was caught up by them at a peak of the Nan-Ling mountain. What Huineng said to them at that time was memorable and is worth mentioning here. Huineng placed the Robe and the Bowl on a rock and said,

> "This robe and bowl are tokens of faith. How can they be taken by force?"

He added,

> "You should put aside all worldly connections. Because you said that you have come for the dharma, do not give rise to even a single thought of idea."

Then the Sixth Master, Huineng, made the final pronouncement:

"Without thinking goodness nor thinking evil, at just this moment, what is your inherent existence?"

This had an awakening impact on the head monk Hui-ming and all the pursuers.

After a month or so, he arrived at a village in Guǎngdōng Province. For several years, Huineng concealed himself among lay people in the Ling-Nan area. In 676, when Huineng was 38 years old, he reappeared dramatically at the temple, Fa-Xing Si (法性寺), i.e. Dharma Nature Temple (also called 制止寺 or 光孝寺), in Guangzhou (広州). At the temple, a dharma lecture on the *Nirvāṇa Sutra* (涅槃経) was given by the dharma teacher Yin-Zong (印宗法師).

It was a windy day, and the temple flags were flapping in the wind. Two monks looking at the flapping flags started an argument. One said the flag was moving, the other said the wind was moving; they argued back and forth, but could not reach a conclusion. Huineng could not withhold his compassion and opened his mouth [57, 67]:

"It is not the wind that moves, it is not the flag that moves; it is your mind that is moving."

His voice gave a great impression to all who heard it. This episode created a deep insight into what the mind we conceive is in everyday life (Fig. 10.18).

More than five hundred years after this event, Zen Master Wu-men Hui-kai (無門慧開, 1183–1260) cited this story and adopted it as Koan No. 29 in his text *Gateless Gate* (無門関) (1228). [67] ([68]). The Master commented on it with his koan spirit as follows:

"It is not the wind that moves; it is not the flag that moves; it is not the mind that moves. Whereabouts can you see what the Master (Huineng) told?." And added, "If they could see it, the two monks got gold when trying to buy iron." [67]

Fig. 10.18   Sketch of the Wind-Flag Hall (風幡堂) at the Fa-xing-si (法性寺, 光孝寺), sketched in 18th CE [80].

It is particularly delightful to learn that the temple Fa-Xing-Si (法性寺) had links to the three important Buddhist monks in the early development of Mahāyāna Buddhism in China. That is, the Fa-Xing-Si was a special temple in Chinese Buddhist history.

Firstly, it was the temple where Gunabhadra constructed the Platform for Buddhist Ordination in the 5th century. He came from South India and arrived at Guangzhou 广州 in 435 CE. Secondly, it is said that Bodhidharma also stayed at this temple while he was in Guangzhou (supposed to be from 524 to 527). The Bowl Washing Well (洗鉢泉,

Fig. 10.19 Sketch of the Bowl Washing Well (洗鉢泉) [80].

Fig. 10.19) is supposed to be proof of Bodhidharma' stay there [80]. It is said that Bodhidharma used it for washing his bowl.

Thirdly, the Sixth Zu, Huineng, was ordained as a bhikkhu monk at the temple Fa-Xing Si (around 676). In the temple, one can see a pagoda where Huineng's hair is enshrined. It is in fact a joy to know that the three great Buddhists had historical links with this temple.

Thereafter, Huineng Dashi (慧能大師 the Sixth Zu) turned the Dharma Wheel at the temple Bao-Lin Si (宝林寺, later renamed Nan-hua Si 南華寺), located in the Cáoxī (曹溪) valley, in the northern town ShaoZhou 韶州 in Guang-Dong (広東省).

10.5.6 *Later explosive development*

There were ten great disciples (Zen masters) in the "East Peak Dharma Gate" of the Fifth Zu, Hongren. Among them, the first was Huineng (慧能 Sixth Zu, the southern school), and the next two were Fa-ru (法如) and Shen-Xiu (神秀 master of the northern school). During this period of extraordinary growth of Zen schools, hot disputes arose between the disciples of Huineng and those of Shen-Xiu over "which is the orthodox school: the southern school or the northern school." This was an inevitable circumstance when the Zen schools expanded explosively, and there arose many schools, not only south and north, but also west, east, central, and so on. However, among the disciples of Huineng, there were also thoughtful groups that were unconcerned with such controversy and interested in the practice of sitting-only dharma. They practiced quietly, seriously, and intimately, without words, and went their own ways.

Nan-yue (南嶽懷讓 677–744) and Qing-yuan (青原行思 660–741) were the two Zen masters who led such serious *sanghas*. Both of them were the disciples of Huineng. As a matter of fact, Chinese Chan originated from the two groups and expanded to larger schools thereafter. Master Shi-tou (石頭希遷, 700–790), a disciple of Qing-yuan, wrote his view on the affairs in the well-known respected verse "Can-Tong-Qi" (参同契):

> The *Mind of a great saint of India,* 竺土大仙心
> *transmitted from west to east without word,* 東西密相付
> *There exist both clever and dull in human beings.* 人根有利鈍
> *In Dharma, there is no teacher of the south nor teacher of the north.*
> 道無南北祖

More than a hundred years later, during the Tang Dynasty (唐朝), the Dharma Lamp of Bodhidharma was respected by emperors and the succeeding masters were bestowed with honorific titles as follows.
The First Master, Bodhidharma, was given the honorific title of
　　Yuan-Jue Dashi (円覚大師 the Great Teacher of Perfect Enlightenment),
The Second Master, Huike, was given the title of
　　Tai-Zu Zen Master (太祖禅師 Great Ancestor Chan Master),
　　and so on.

After the Sixth Zu, Huineng, the Dharma Lamp of Bodhidharma was succeeded by his two disciples, Nan-yue and Qing-yuan, from whom five (or seven) Zen schools sprang and flourished in later generations. The five major schools are

Lin-Ji school (臨済宗); Gui-Yang school (潙仰宗): [lineage of the Seventh Zu, Nan-Yue], Cao-Dong school (曹洞宗); Yunmen school (雲門宗); Fa-Yan school (法眼宗): [lineage of the Seventh Zu, Qing-yuan].

Among the five major schools, three schools (Soto school 曹洞宗, Lin-zai school 臨済宗, and O-baku school 黄檗宗) were succeeded in Japan.

Even in the twenty-first century, the Dharma Lamp of Bodhidharma shines brighter than ever, appealing to people and providing them with islands of peace of mind. Finally, the author wishes to emphasize the depth and scale of Bodhidharma's influence on the East and the world. Among hundreds of koans of the Chan school (in Chinese), the most frequently asked question is, "What is the intention of Bodhidharma coming from the West-land (India)?" This question was first asked by the Second Zu, Huike (二祖慧可大師) (see Section 10.2). The question

"Why did Bodhidharma come to the East land?"

is also the koan to the people living in the present world of the 21th century.

10.6 Zen Buddhism in Japan

10.6.1 *Legacy of Zen Buddhism*

The Mahāyāna Buddhism of Chan School (禅宗) brought about revolutionary change in East Asia and enlightenment to both culture and the daily lives of people there. The Chinese Chan (禅) is called Zen in Japanese. Japanese societies learned their lifestyles from Zen monasteries or Buddhist temples. Zen Buddhism has succeeded up to the present day, in the 21st century.

According to a stone monument [78] erected in AD 689 in China, "The dharma teacher Bodhidharma brought the dharma of Shakyamuni to the east land from India, and came to the country Wei (China), and transmitted it to Huike (Second Master). And so on " (Section 10.5.1). From this lineage, several schools of Zen Buddhism sprang out and flourished in later generations. *Zen* (禅)' is the first letter of the phonetic translation of the Sanskrit *dhyana* (禅定). The philosophy of Bodhidharma is expressed compactly as

"Special transmission outside the scriptures, directly pointing to the mind dharma."

Among the several Zen lineages in China, three (Lin-zai school 臨済, Soto school 曹洞宗 and Oh-baku school 黄檗宗) were transmitted to Japan from the 12th century on. The philosophy and practice of Zen Buddhism had a profound influence on forming the core of Japanese modern culture, combined with traditional Japanese cultures.

Zen Buddhism is not a simple religion for worship, but it is essentially connected to daily practice. Beyond any dogmatic concept, it is based on the *śūnyatā* (*emptiness*) philosophy. Any Buddhist activity should not be detached from daily life. It is natural that kitchen work is regarded as an essential practice of monks at Zen monasteries, where *kitchen worker* is regarded as an honorary position because these workers can support all the practitioners in the monastery from an essential standpoint (Fig. 10.20).

From the 7th century on, envoy ships were sent from Japan to China to learn from China. In particular, Zen Buddhism was brought to Japan in the 12th century (partly, from 9th). This was based on *mind transmissions* from Chinese Chan masters to Japanese Zen masters, carried out from time to time via various routes. There were several attempts to establish Zen schools in Japan.

Zen Master Eisai (栄西禅師) is usually credited with the first transmission of the Rinzai (臨済 Lin-Ji) Zen school to Japan after his return from China in 1191. The Rinzai school is the Japanese lineage of the Chinese Lin-Ji school of the Chan Master Lin-Ji (臨済) in the Tang Dynasty. Later, Daito Kokushi (大燈国師 1282–1338) and Musō Kokushi

(a)   (b)

Fig. 10.20 Monks at a Zen monastery: (a) Kitchen work. (b) Tea time.

Fig. 10.21 Stone (dry) garden at Ryōan-ji Zen temple (Kyoto) [85], belonging to Myoshin-ji, Rinzai school.

(夢窓国師 1275–1351) (Kokushi 国師: national teacher) in Kyoto are said to have achieved a distinct identity that is characteristic of Japan.

The times during which the Rinzai Zen school was established in Japan coincided with the rise to power of the samurai class (warriors) in the country. Along with early imperial support, this school had the patronage of this newly ascending warrior class. This collaboration resulted in the creation of remarkable gardening art. One of the well-known examples is the stone (dry) garden in Fig. 10.21.

It was 1223 CE when a young Japanese monk, Do-Gen (道元), crossed the sea together with his senior colleague monk, wishing to learn authentic Buddhism. They visited various temples in China to find a true Buddhist teacher. Finally, the monk Do-Gen visited the Tian-Tong Temple (天童寺), south of Shang-hai, and met a Chan master, Ru-Jing (如浄禅師). Because Master Ru-Jing was a true teacher, Do-Gen asked Master Ru-Jing (如浄禅師) various questions in order to ensure his understanding about Buddhism. He recorded their dialogues in his private note [84] (discovered after his nirvāṇa). One question was as follows.

"Why is the main aim of Bodhidharma's visit to China is said to be 'special transmission outside the scriptures'?"?

The Master responded:

"Before Bodhidharma, only envoys and furniture had come to China, but no master was there. When Bodhidharma came to China, it was as if the people had a king."

This was the understanding of the Chinese master after seven hundred years of Bodhidharma's time. The monk Do-Gen received the dharma of the twenty-third Zu of Chan Buddhism, Ru-Jing (第 23 祖如浄禅師, 50th from Mahākāśyapa of India). In this way, Zen Buddhism was transmitted to Japan.

We later generations are benefiting endlessly as the descendants of Two Bodhisattvas: the First Zu, Bodhidharma the Great Teacher, and the Second Zu, Huike the Great Disciple.
Moreover, the benefit is open to everyone in the world.

10.7 Dhyana and Mind Are Not Separate

Zen Monk Do-Gen founded the Japan Soto school after returning to Japan in 1227. Later, in 1243, he built the Eihei-ji Monastery 永平寺 (Fig. 10.22) on a mountain slope in central Japan, where they have heavy snow every winter, and founded the Japan Soto school (日本曹洞宗). Following the

(a)   (b)

Fig. 10.22 (a) Eihei-ji Monastery 永平寺 the photo taken by the author in 2012. (b) Dhyana meditation in a monk hall. Eihei-ji Monastery 永平寺, Fukui province, Japan [85].

advice of Ru-Jing Dashi, the place was chosen far away from the capital, Kyoto. Even in the 21st century, a number of monks practice the zazen sitting practice there. The formality and details of every action in the monastery are observed strictly.

The style shown in Fig. 10.22(b) is not a simple formality at all, but rather follows the authentic tradition of Buddhism. When a session starts, all monks usually chant,

大我解脱服 霜枯福田衣 披零細乗鞍 広虔議衆生

Great-great vestment of liberation;
Form-less robe blessed with potential merit like rice-field,
Respectfully, wrapping oneself with Tathāgata Dharma,
Broadly saving all living beings like a ferry with compassion

This resonates with the *Xin-Xin-Ming*, praising *Non-duality of Practice and Mind*. The *Dhyana* is practiced with the Padmasana sitting posture with the vestment robe representing Faith. *The true world of Mind* emerges resonantly with such *Dhyana*, transcending words and scriptures. The *Xin-Xin-Ming* concludes by the verse:

Dhyana and Mind are non-dual.   *Non-dual Dhyana-Mind.*

In fact, already in 414 CE, the Chinese Monk *Fa-xian* recorded on the *dignified demeanor* of Buddhist monks in cloud-like gathering. After returning from fifteen years of travel to and from India, he wrote his great travel record 法顕伝 (*Record of Buddhistic Kingdoms*) [83]. At the end of his travel, he recollected his great journey and wrote compactly about the fresh impression he had received from the Buddhists he met on his way to India in the sandy countries of central Asia. He wrote as follows.

> *From the sandy Gobi Desert on, westwards to India, the beauty of the dignified demeanor of Buddhist monks and the transcendency of the Buddha dharma were so admirable and beyond the power of language. Its full description with words is impossible.*

Buddhists try to realize the transcendence of the dharma, which is beyond the power of language.

Formality of the body-fixation is concerned with its external appearance, which is accompanied by the internal peace and depth of *samādhi*. Bodhidharma said to his disciples:

> *"Dharma is beyond language description. Dharma of sitting gives us fearlessness externally. It gives us supreme peace internally. There is no difference between common man and sage in this respect."* (Section 10.3.6 (iv))

What the zazen sitting looks like can be seen in the photos of Fig. 10.23 [86]. From the outside, the practitioner appears to be sitting still but is actually moving slowly for the need of deep breathing. Not only

Fig. 10.23   Formality of zazen sitting in the Japan Soto Zen school [86].

does the practitioner's body move when breathing, but the heart never stops beating for blood circulation, which is the most essential function in a living body.

In other words, the internal physiological mechanisms are continuously evolving and functioning autonomously within the body under deep breathing and blood circulation. This is the essential aspect of dhyana. In fact, before beginning the dhyana practice, the human body is disturbed by external stresses. Once the practitioner begins to take the dhyana posture in a quiet environment, external perturbing stresses disappear and the state of being disturbed subsides. After a while (about 15 minutes or so), the body state is freed from external disturbances and is liberated. From this moment, the innate autonomous body transformation mechanism begins to govern the practitioner's body physiologically. Taking the dhyana posture, the practitioner looks unmoved externally except for the faint motion of deep breathing, but internally the innate mechanism of transformation works autonomously and freely. Functioning at full capacity, the internal physiological mechanisms enjoy a stress-free state, with clear awareness not only of the inside but also of the outside, without a fixed target, as if floating in space. This is, as it were, the dhyana of high state, which is comfortable, serene, and peaceful.

10.8 Nirvāṇa Does Not Have the Slightest Difference From Samsara

In the *Madhyamaka-kārikā*, Acārya Nagarjuna stated,

> *The samsara does not have the slightest difference from nirvāṇa.*
> *Nirvāṇa does not have the slightest difference from the samsara.*

(Section 9.5, (22))

Seeing the photo (Fig. 10.24) below, the present author is strongly struck by the scene of the changing world, as if the body sways and eyes turn. Is it *nirvāṇa* or *samsara*?

In 642 CE, fifty years after the misfortune of the Second Zu, Huike Dashi (Section 10.5.3), the Yuen-fu Temple was constructed at his last

Fig. 10.24   Pagoda re-erected in 1057 CE at the Yuen-Fu Temple (元符寺), where the relics of Hui-ke Da-shi were enshrined. The photo was taken in 1945. The location is at He-bei Province (河北省邯郸市成安县二祖村): https://kknews.cc/fo/85p39pl.html

resting place in Hebei Province (China) in memory of the Dashi. Sacred relics of the Great Zu Huike were enshrined in the pagoda constructed in 732 CE. The pagoda in Fig. 10.24 (the photograph of which was taken in 1945) was re-erected in 1057 CE [79]. When the author went there in 2009, he could not find the pagoda. According to a Chinese news site [79], it had been completely destroyed in 1969 during the Cultural Revolution, but now the temple is being rebuilt.

This is certainly **samsara**, which in Sanskrit means "world," or "transmigration," or *"passing through a succession of states."* However, when all views have been cast off, *emptiness* and *nirvāṇa* emerge immediately, because all views based on historical records are intrinsically empty. The first pagoda constructed in 732, had probably been standing in a dignified appearance for about three hundred years. Then the pagoda reconstructed in 1057, perhaps rebuilt, had been standing gloriously and looked like the photograph in Fig. 10.24 in 1945, after about nine hundred years. Finally, it was destroyed in 1969, about fifty years ago. It is certainly the *nirvāṇa* of a pagoda.

When one looks at the historical descriptions of pagodas, as described above and connects those words or ideas to them, one can be sure that this world is *samsara*. However, without such words conveying a *sense of sentiment*, the whole sequence concerning the life of two pagodas is nothing but *nirvāṇa*. There does not exist the slightest difference between *samsara* and *nirvāṇa*. Wording is a merely verbal description, equivalent to saying that it is *empty*. One cannot describe *nirvāṇa* with words.

The *nirvāṇa* world is *ineffable* and *incomprehensible*; it cannot be understood with language, as hinted previously in Section 10.7. Observing the historical revival of Buddhism in the modern world, one is struck by the story and the memory of Huike Dashi. In fact, our world is evolving on its own, serenely and solemnly, without a word:

The world is *nirvāṇa*, which *does not have the slightest difference from samsara.*

Chapter 11

Modern Sciences: Natural World and Dependent Origination

Twin typhoons over the Pacific Ocean (June 29, 2004). By courtesy of Japan Weather Association.

11.1 Goal of Science

In his celebrated book *To Explain the World* [88], Steven Weinberg, an American theoretical physicist and Nobel laureate, mentioned modern physical sciences. What is most important is as follows:

"Science could hardly exist without activities of human beings."

Science has been discovered in order to explain nature. Modern science is an approach as well as a technique sufficiently well tuned to nature so that they work efficiently, and the approach is a practice that allows us to learn reliable things about the world. The development of agriculture can be taken as an example to understand the development of science. Agriculture is the way it is because its practices are sufficiently well tuned to the realities of biological objects (flora and fauna) so that it works appropriately.

The goal of science is to find explanations of natural phenomena that are purely naturalistic. Science is cumulative. Each new theory proposed at a time incorporates successful earlier theories as approximations and explains why those approximations worked well. This is an *explanation,* quite different from a simple description. Validity of the theories should be *verified* finally by experimental tests of their predictions. In this way, scientific activities are considered to be endless, namely scientific activity is a sequence of proposals and verifications in order to approach the reality unlimitedly.

The present form of *modern science* is qualitatively different from what it was at its start in the ancient Greek world, more than two thousand years ago. An essential element of modern science is **verification by experimental tests** or accurate observations of nature. None of this was obvious to the ancient worlds, with a few notable exceptions.

Physics and astronomy underwent revolutionary changes in the sixteenth and seventeenth centuries in European countries, which led to their modern form, providing a paradigm for the future development of all sciences. A historian described this change by the term *scientific revolution* and stated that it outshines everything since the rise of Christianity two thousand years ago. With the perspective of modern science, it was a real discontinuous change in intellectual history (from Part IV of [88]). Science has nothing to say about the existence of God or the afterlife.

11.2 Science Exchange Between East and West in the BCE Ages

During the Greek invasion of the Persian Empire in 334–324 BCE, a person (Pyrrho) from the Greek Elis visited the northern part of India together with Alexander the Great. There he met early Buddhist masters. In Chapter 1, we saw a historical dialogue at a later era which was another encounter between an Indo-Greek king and an Indian Buddhist, resulting in the dialogue Sutra, *Milinda Pañha*, between the Buddhist sage Nâgasena and Indo-Greek king Menander I of Bactria in the 2nd century BCE. Bactria was located in central Asia, which covered modern-day Afghanistan, Uzbekistan, and Tajikistan and extended, to the westward, up to the southern side of the Caspian Sea. With these encounters, Indian philosophy and early Buddhism were influenced *by* and conversely communicated *to* ancient Greece, such as a form of Pyrrhonian philosophy in Greece [16]. In this regard, one can make the following comparison between the west and the east.

In ancient Greece, a classic view established by Empedocles in the 400s BCE [88] was that all matter is composed of four elements: *earth*, *water*, *fire*, and *air* (or *wind*). However, Democritus (c. 460–370 BCE), who was born at Miletus in Anatolia (known as Asia Minor too in present-day Turkey) and settled in Abdera, an ancient *polis* in the northeast of Greece, believed that all matter consists of tiny invisible particles called *atoms*, which move in empty space:

> *Sweet exists by convention, bitter by convention; atoms and empty space exist in reality.*

Like modern scientists, early Greeks were interested in looking beneath the outward appearance of matter and pursued the knowledge about a deeper level of reality. At first glance, it does not appear that all matter is made of water, or air, or earth, or fire, or all four together, or even of atoms.

One can see some influence of the Greek natural philosophy on the Indian philosophy. The following description is given in the first chapter of the *Abhidharmakośa-bhāṣya* (*AKB*) [2] (Chapter 2 of the present book) by Vasubandhu (late 4th CE common era):

> *What is the intrinsic nature of the four elementary substances?*: (i) *Earth*, (ii) *water*, (iii) *fire*, (iv) *wind are, in their intrinsic natures,* (1) *solidity,* (2) *humidity,* (3) *heat,* (4) *mobility,* respectively.

In the Buddhist *abhidharma* interpretation, these four elements are called *Four Great Elements*, or *Four Fundamental Material Elements*.

Remarkably, they coincide with the four elements established by the Greek philosopher *Empedocles*.

Furthermore, there is an *atomic* interpretation in Chapter 2 of the *Abhidharmakośa-bhāṣya*:

> By *paramanu* (an atom in *Sanskrit*), *one does not understand the atom here in the strict sense, the atom which is a single real entity, but rather the composite molecule."* In addition, *"the composite molecule includes eight real entities: namely the four fundamental material elements and the four derivative elements, namely visible form, odor, taste, tangible.*

Chapter 1 contains the following interpretations:

> *atom as a monad would not be a material form.*
> *A single material form as a monad never exists in an isolated state. However, in the state of a composite, it is susceptible to deterioration and to offering resistance body.*

Since the *Abhidharmakośa-bhāṣya* was composed in the late 4th century CE, it is very likely that the atomic interpretation was an influence of the Greeks' natural philosophy.

11.3 Ancient Greek Natural Philosophy

(*i*) *Did early Greek thinkers attempt to test their speculations*?
It is true that there are some notable achievements in ancient Greek natural philosophy, which are still valid and fundamental. However, we must admit an important aspect that is almost missing in all Greek thinkers of natural philosophy. None of them attempted seriously to justify their speculations. This is just as true of Democritus as of the others. We do not see any effort in Democritus' books to show that matter really is composed of atoms. Steven Weinberg reviews this aspect very critically in his book [88] and writes in Chapter 1:

> *The early Greeks had very little in common with today's physicists. Their theory had no bite. Empedocles could speculate about the*

*elements, and Democritus about atoms, but their speculations led to no new information about matter — and certainly to nothing that would allow their theories to be **tested**. It seems to me that to understand their early Greeks, it is better to think of them not as physicists or scientists or even philosophers, but as poets.*

Plato (lived before and after 400 BCE) was a philosopher and the founder of the Academy in Athens, the first institution of higher learning in the Western world. It seems at times that Plato did not intend to be taken literally. As an example, he introduced the story of *Atlantis*, which supposedly flourished thousands of years before his own times. Nobody can tell that he really knew anything about what had happened thousands of years earlier.

Thales (lived in Miletus for about 78 years before and after 600 BCE), who lived about two centuries before Plato, is regarded as the first philosopher in the Greek tradition. Thales was said to believe that all matter is composed of a single fundamental substance. Much later, a historian wrote, "His doctrine was that water is the universal primary substance, and that the world is animate and full of divinities."

(cited from Chapter 1 of [88]).

From Thales to Plato, all Greek philosophers of nature lack the insistence to verify whether the theory actually applies to the real world, or willingness to keep an open mind regarding their predictions to be tested experimentally. Present-day physicists test their speculations about nature by using proposed theories to draw more or less precise conclusions that can be tested by observation. This did not occur to the early Greeks, or to many of their successors, for a very simple reason: *They had never seen it done* [88].

(ii) Some notable achievements in ancient Greece and Egypt

We know that there were many notable achievements or observations by thinkers in ancient Greek and Egypt (Alexandria) or others:

- ✧ Pythagoras (c. 570–495 BCE) founded the so-called Pythagorean school (cult). He discovered arithmetic rules in musical tones played by stringed instruments: If one string is two-thirds the length of the other, the two strings produce a particularly pleasing chord. A great contribution by Pythagoras was the Pythagoras theorem, a fundamental relation in Euclidean geometry among the three sides of a right triangle. The theorem states that the square of the hypotenuse, c, is equal to the sum of the squares of the other two sides, a and b. The equation can be written as follows:

✧ *Aristotle* $a^2 + b^2 = c^2$. (384–322 BCE) cited empirical evidence for the spherical shape of the Earth, such as the following: (1) the Earth's shadow on the moon during a lunar eclipse is curved; (2) the position of stars in the sky seems to change as we travel north or south. He understood that the shape of Earth is round.

✧ Most impressive scientist of the *Hellenistic* era (323–30 BCE) was *Archimedes*. His greatest achievement was *Archimedes' principle* about floating bodies, which states that a solid body immersed in a liquid is lighter than its true weight by the weight of the liquid displaced by the body.

✧ In addition to *Archimedes*, a great Hellenistic mathematician was a younger contemporary *Appollonius*. One of his works was on conic sections: the ellipse, parabola, and hyperbola. These are curves that can be formed by a plane slicing through a cone at various angles. Much later, the theory of conic sections was crucially important to Kepler and Newton, but it found no physical applications in the ancient world.

(*iii*) How modern science began

The aforementioned thinkers were certainly brilliant, but with emphasis on *geometry* only. In fact, *algebraic* techniques, essential in modern physical science, were missing from Greek mathematics. The Greeks never learned to write algebraic formulas. Formulas such as $E = mc^2$ and $F = ma$ or *differential* equations (developed by L. Euler) are at the heart of modern physics.

However, the geometric approach persisted until well into the scientific revolution of the *seventeenth* century. In his book, Galileo compared the universe to an all-encompassing book and praised mathematics for enabling one to read it: "It cannot be understood unless one first learns to understand the language and knows the characters in which it is written. It is written in mathematical language, and its characters are triangles, circles, and other geometrical figures." Geometry had been the tool until the times of Kepler, Galileo, and Newton. Even in the masterpiece *Principia*, Newton used geometry to propose his revolutionary approach to mark a turning point.

From the time of flourished Greek philosophy to the time of modern physical science, there existed a large separation of nearly two thousand years. Scholars of modern physical science had to invent powerful tools to defend themselves and justify their speculations by means of experiments, accurate observations, and mathematics. During the initial periods of emerging modern physical science, it had to experience hard times. Scientists had to confront opposing reactions from the authorities of existing Christianity.

11.4 Dawn of Modern Physical Science

The revolutionary change of physics and astronomy that happened from the sixteenth to eighteenth centuries was a real turning point of history of human intellectual activity, after which these sciences assumed their modern form.

(i)   Observation of celestial bodies and Kepler's laws

Copernicus (1473–1543) and Kepler (1571–1630) proposed *heliocentrism*: It is the Earth, not the Sun, that moves in the solar system. In other words, the Earth and planets revolve around the Sun at the center of the Solar System. Why did they propose *heliocentrism*? The reason was not its better agreement with observation (than the earlier geocentric model of the Ptolemaic system) but the simplicity and coherence of the theory of *heliocentrism*. However, their proposal of the Copernican system invited opposing reactions from religious leaders [88, Chapter 11]. A few years later after the publication of *De Revolutionibus* (1543) by *Nicolaus Copernicus*, a colleague of Martin Luther joined the attack on Copernicus, citing a passage from the *Ecclesiastes* (one of the Writings of the Bible): *"The Sun also rises, and the Sun goes down, and hastens to his place where he rose."* Conflicts with the literal text of the *Bible* would naturally raise problems for *Protestantism*, which had replaced the authority of the Pope with that of Scripture. Beyond this, there was a potential problem for all religions: Man's home (the Earth) had been demoted to just one more planet among the other five. Tycho-Brahe was a most proficient astronomical observer (born in 1546 as a Danish nobleman). When he passed away in 1601, he left his unprecedently accurate data of his observations of the motion of Mars. Tycho's data were so good that Kepler could see some inconsistency between the observation and theory, where the theory was based on the assumption that Mars' orbit is a circle around the Sun at its center. At some point, Kepler became convinced that he had to abandon the assumption. Thus, owing to Kepler's precise calculation and insight, there emerged a correct description of the solar system, encoded in Kepler's three laws:

I. Planets move on ellipses, with the Sun at the focus, not at the center.
II. As a planet moves, the line between the Sun and the planet sweeps out equal areas in equal times.
III. The ratio that exists between the periodic times (T) of any two planets is precisely the ratio of 3/2-power of the mean radii (a). Namely, T^2 is proportional to a^3.

Kepler's laws gained observational support from Galileo's (1564–1642) many astronomical discoveries using his telescope, which include (1) Jupiter's moons; (2) that the surface of the Moon is uneven and full of depressions and bulges (1609) like the face of the Earth; (3) and seeing the planets through his telescope as *exactly circular globes that appear as little moons* (1610). In fact, Galileo's discovery gave important support to the Copernican theory. First, the system of Jupiter and its moons provided observational evidence implying the reality of heliocentrism proposed by Copernicus, a celestial system of planets in motion around the Sun. In addition, the example of Jupiter's moons put to rest the objection to Copernicus that if the Earth is moving, why is the Moon not left behind? Everyone agreed that Jupiter is moving, and yet its moons were evidently not being left behind.

When Copernicus was able to explain the old theory by his new view of the Earth moving around the Sun, what a pleasure he must have felt! How much the mathematically gifted Kepler must have enjoyed replacing the (still flawed) Copernicus theory with the planetary motion in ellipses, obeying his three laws!

(ii) Studies by laboratory experiment

• *Galileo did the first experimental study of motion*

Experimental study of the motion of a body began with Galileo, which marked the real beginning of modern experimental physics. Furthermore, his quantitative study of the motion of falling bodies was an essential prerequisite to Newton's subsequent work. Galileo conducted an experiment on a falling body. The speed of a ball falling down vertically changes continually; namely it is *accelerating* downward, which is caused by the Earth. In another experiment, ball was rolled down an inclined plane from its initial static state. Galileo found that its downward speed changes continually and increases in proportion to the square of the time elapsed.

Galileo's another experiment is described in *Dialogues Concerning Two New Sciences* (1635). A ball is allowed to roll down an inclined plane. It then rolls along the horizontal tabletop on which the inclined plane sits and finally shoots off into the air from the table edge. By measuring the timewise motion of the ball and observing the trajectory of the ball in the air, Galileo concluded that the trajectory is a *parabola*. This was a historical triumph of Galileo's experimental study.

However, as fate would have it, he had to face a serious social problem as a result of his unprecedented astronomical discoveries. In 1615, Galileo went to Rome to argue against the suppression of *Copernicanism*. The Pope decided to submit the Copernican theory to a panel of theologians, whose verdict was that the Copernican system is foolish and absurd in philosophy, and formally heretical inasmuch as it contradicts the express position of Holy Script in many places. Galileo was summoned in 1616 to the inquisition and received two confidential orders, ordering him not to hold, defend, or teach *Copernicanism* in any way [87, Chapter 11].

• *Huygens studied accelerating motions*

Velocity of a body is defined by the speed and direction of its motion. Such a quantity is called a *vector* in physics. *Acceleration* is also a directed quantity (a vector) with its magnitude defined by the change of speed per time elapsed. When a body is moving along a straight line with a constant speed, its acceleration is zero. But there is an acceleration in circular motion even moving at a constant speed because its direction changes continuously — it is the *centripetal* acceleration, consisting of a continual turning toward the center of the circle.

In 1656, Huygens (1629–1695) invented a pendulum clock. The clock was based on Galileo's observation that the time a pendulum takes for each swing is independent of the swing's amplitude (Galileo recognized the property in the limit of very small swings). The motion of an oscillating pendulum consists of the acceleration along a circular section as well as falling by gravity. Huygens was able to show that *the time* (T/2) for a pendulum of length/to swing through a small angle from one side to the other (half of the oscillation period T) equals a constant π times *the time* $\sqrt{l/g}$ for a body to fall a distance equal to half the length $l/2$ of the pendulum. Using this principle, Huygens was able to calculate the acceleration g due to gravity, which Galileo could not measure accurately. This fact was of great importance to Newton's work.

Huygens (and Newton) concluded that a body of mass m moving at a constant speed v around a circle of radius r is accelerating toward the center of the circle, with the acceleration v^2/r; so the force f needed to keep it moving on the circle (rather than flying off in a straight line into space) is $m(v^2/r)$. Consider a weight m placed at the end of a *cord* swinging around a circle. The resisting force to this centripetal acceleration is experienced as the centrifugal force. For weight m, the centrifugal force is resisted by tension in the *cord*. But planets are not attracted by cords to the Sun.

What is it that resists the centrifugal force produced by a planet's nearly circular motion around the Sun? The answer to this question led to Newton's discovery of the inverse square law of *gravitational* force.

(iii) **Newtonian synthesis and dawn of modern science**
Newton (1642–1727) crossed the frontier between the past natural philosophy and what became modern science (as neither a medieval scholar nor an entirely modern scientist). He tied up, innovatively, strands of physics, astronomy, and mathematics. It was he who disclosed the background connections among them, which perplexed philosophers since Plato. Newton was born in 1642 and achieved a number of great works in his whole life, which were mainly after 1664.

Newton's greatest historical impact were the theories of motion and gravitation. It is not an easy guess that the motion of an object that falls toward the ground is caused by a certain force called gravity. More than that, it is far from obvious to imagine that there is any connection of gravity with the orbiting motion of planets. According to Newton's recollection after fifty years (in the 1710s), it was in 1666:

> *I began to think of gravity extending to the orbit of the Moon. From the Kepler's rule connecting the periodical times of the planets with their distances from the center of their orbits, I deduced that the forces which keep the planets in their orbits must be reciprocally as the squares of their distances from the centers about which they revolve.* **Thereby I compared the Moon in her orbit with the force of gravity at the surface of the Earth. (The force acting on the Moon was compared with the force at the Earth's surface.) I found them answer pretty nearly.**
>
> *All were in the two* plague *years of 1665–1666. For in those days, I was in the prime of my age for invention and minded Mathematics and Philosophy more than at any time since.*

(In 1665, the plague (pest) struck Cambridge, the university largely shut down, and Newton went back home to Woolsthorpe. In 1656, Huygens found the circular motion of a body to be an accelerating motion even moving at a constant speed.)

(iv) **Mathematical Principles of Natural Philosophy (Principia** in short, published in 1687)

A modern physicist may be surprised to see how little Newton's masterpiece "*Principia*" resembles any of today's treatises on physics. There are many geometrical diagrams, but few equations. They had to wait until the next century to see expressions of algebraic equations or differential equations used nowadays in mechanics.

In *Principia*, Newton proposed the famous three laws of motion:

Law I: Every body preserves its state of being at rest or of moving uniformly straight forward, except insofar as it is compelled to change its state by forces impressed.

Law II: A change in motion is proportional to the motive force impressed and takes place along the straight line in which that force is impressed.

Law III: To any action there is always an opposite and equal reaction. In other words, the actions of two bodies upon each other are always equal, and always opposite in direction.

In addition, summarizing the evidence from astronomy, Newton concluded that *gravity* exists in all bodies universally and that it is a central force that goes as the inverse square r^2 of the distance r between two interacting bodies and is directed along the line connecting the two bodies. The term "universally" means that the accelerations produced by gravitation are independent of the nature of the body being accelerated, whether it is a planet or a moon or an apple. With this gravitational force, Newton could account for all of three Kepler's laws.

The gravitational force F between two bodies, of mass m_1 and mass m_2, separated by a distance r is represented by $F = G\, m_1\, m_2/r^2$, where G is a universal constant. Neither this formula nor the constant G appears in Newton's treatise *Principia*. Those expressions improved later.

Newton proved that central forces (forces directed toward a single central point) and only central forces give a body a motion satisfying Kepler's second law, and that (only) central forces proportional to the inverse square r^{-2} of the distance (like gravity) produce motion orbiting on a *circle*, an *ellipse*, a *parabola*, or a *hyperbola*.

In 1684, Newton received a visit from the astronomer E. Halley in Cambridge, who had already seen the connection between the inverse square law of gravitation and Kepler's third law for circular orbits. Also, both Hooke and Wren had the same view. It is said that Ismael Bullialdus, a French priest, was the first to suggest this connection, in 1645, which was quoted by Newton later.

(v) **Newton did not try to explain why gravity exists**
Newton wrote:

> *Thus far, I have explained the phenomena of heavens by the force of gravity, but I have not yet assigned a cause of gravity. Indeed, this force arises from some cause that penetrates as far as the centers of the Sun and planets without any diminution of its power to act, and that acts not in proportion to the surface areas of the particles on which it acts, but in proportion to the quantity of solid matter, and whose action is*

extended to immense distances, always decreasing as the inverse squares of the distances.

*I have not as yet been able to deduce from phenomena the reasons for these properties of gravity, and I do not "**feign**" hypothesis.*

Newton was not able to explain why gravity existed. He believed that gravity existed and was a physical object quite different from what he knew.

His theory of gravitation was opposed in France and Germany. The opponents argued that an attraction operating over millions of miles of empty space would be an occult element in natural philosophy, and they further insisted that the action of gravity should be given a rational explanation, not merely assumed.

In doing this, natural philosophers on the continent were sticking to an old ideal for science, going back to the Hellenic age, that scientific theories should ultimately be founded solely on reason. However, we have learned to give this up. Even though our successful theory of electrons and light can be deduced from modern standard model of elementary particles, which may in turn eventually be deduced from a deeper theory, however far we go, **we will never come to a foundation based on pure reason.**

Most physicists today are resigned to the fact that we will always have to wonder why our deepest theories are not something different. Namely, we will never come to an end on pure reason. Experimental evidence will decide which speculation is closer to the reality. Modern science is impersonal, without room for supernatural intervention or human values; it has no sense of purpose; it offers no hope for certainty. In addition, Newton stated that "I have not as yet been able to deduce from phenomena the reasons for these properties of gravity and I do not feign hypothesis."

It is interesting to see some similarity between these observations and what Buddha told his disciples when he was asked whether the world is infinite or finite, or whether the world is eternal or non-eternal. We will come back to this point later again.

11.5 What is Modern Science — 1

The present form of modern science is qualitatively different from what it was during the dawn of the modern science of the sixteenth century.

In fact, geometric approach had persisted until well into the seventeenth century. Essential elements of modern science are precise experimental tests, accurate observations of nature, and mathematically founded theories giving predictions that can be tested experimentally or observationally. The revolutionary change has been made in physics and astronomy in the sixteenth and eighteenth centuries. A historian described this change as *scientific revolution* and stated that it outshines everything since the rise of Christianity. From the perspective of modern science, it was a real change in intellectual history (from Part IV of [88]). Science has nothing to say about the existence of God or an afterlife. In Western societies, there are two main streams of science and religions.

(a) *Concept of physical fields*
The foundation of classical physics had been laid by Isaac Newton. Building on the discoveries by Kepler, Galileo, Huygens, and others, Newton developed laws that described a comprehensible mechanical universe: Planets are moving on elliptic orbits around the Sun located at one of the focal points of the ellipses; an apple falling to the ground and an orbiting moon are governed by the same law of gravity, and theoretically the motion of a particle with mass could be explained, determined, and predicted by geometrical approaches. Mechanics described arithmetically and analytically by differential equations were developed by L. Euler (1707–1783) and J.-L. Lagrange (1736–1813). In particular, they described Newton's equation of motion by a second-order differential equation with respect to the time t.

A new physical concept, i.e. the field, appeared in the 1800s, which was the most important invention since Newton's time. Specifically, Michael Faraday (1791–1867) discovered the electric field and the magnetic field. It took great scientific imagination to realize that the *field* in the space between the charges and the particles, not the charges or the particles themselves, is essential for the description of physical phenomena. The field concept proved successful when it led to a system of partial differential equations, formulated by Clerk Maxwell (1831–1879), that described electromagnetic fields. From the equations, Maxwell derived that oscillating waves of electromagnetic fields propagate with the speed of light, *c*. This result led to a new important discovery that light is nothing but an electromagnetic wave.

(b) *Modern Prometheus*

Science has nothing to say about the existence of God or an afterlife. The term *"Modern Prometheus"* was actually coined by Immanuel Kant in reference to Benjamin Franklin, who conducted an experiment in which he flew a kite during a thunderstorm to extract electric sparks from clouds in Philadelphia (1752, USA). A novelist Mary Shelley (1797–1851) saw Prometheus not as a hero but rather as something of a devil and blamed him for bringing fire to man and thereby seducing the human race to the vice of eating meat (fire brought cooking and further brought hunting and killing). Mary Shelley is known for her novel *"Frankenstein"* or, *"The Modern Prometheus"* (1818), which was an early science fiction.

The story begins in the Arctic Sea: *"In 1794, Captain Walton leads a troubled expedition to reach the North Pole. While their ship is trapped in the ice of the Arctic Sea, the crew discovered a man,* **Victor Frankenstein,** *traveling across the Arctic on his own."* This is an initial plot of the movie *Mary Shelley's Frankenstein* (1994). The rescued Frankenstein told the skeptical captain the ghastly story of how he created a living monster out of exhumed corpses. Mary is likely to have acquired some ideas for Frankenstein's character from the book *Elements of Chemical Philosophy*, which describes that *"science has bestowed upon man powers which may be called creative; which have enabled him to change and modify the beings around him."*

One of the most notable natural philosophers among Shelley's contemporaries was Giovanni Aldini (1762–1834), who made many public attempts at human re-animation through bio-electric galvanism in London. He was an Italian physicist and became professor of physics at Bologna in 1798. His scientific work was chiefly concerned with galvanism, anatomy, and its medical applications. In biology, *galvanism* is the contraction of a muscle that is stimulated by an electric current. In physics and chemistry, it is the induction of electrical current from a chemical reaction, typically between two chemicals with different electronegativities. The modern study of galvanic effects in biology is called electrophysiology, the term galvanism having been used in historical contexts only. The term is also used to refer to bringing organisms to life using electricity, as popularly associated with Mary Shelley's work *Frankenstein*. People still speak of being "galvanized into action," meaning getting stimulated into some activity. Giovanni Aldini performed a famous *public demonstration* of electro-stimulation of the limbs of an executed criminal at Newgate in London in 1803. The movie *Mary Shelley's Frankenstein* (1994) was a tale of classic horror.

Science is a *double-edged sword*. Although it has a number of glorious stories of success, it is also accompanied by grave anxieties. Mary Shelley's Frankenstein hints at the other aspect of science.

(c) *Modern Prometheus is not a horror from the view of emptiness*

Modern Prometheus might be viewed differently from Buddhists' eyes. In Chapter 9, we saw how Buddhists hold their view on the actions of everyday life, learning from the verse of Nāgārjuna in Section 9.5 (8). The sequence of a karmic action, the acting agent (person), and its consequence is understood as a sequence of phenomena of emptiness. Consider a man possessing magical powers, who creates an illusory body, and suppose that the illusory body creates another illusory body. In that way, we can view ourselves, our karmic action, and the consequence of our karmic action. The agent of action (we) is like the first illusory body. Its karmic action is like the illusion's illusion. Buddhists hold their view as follows. Afflictions, karmic actions, agents, and resulting fruits are all like a sequence of mirages or dreams. All of them lack intrinsic nature. All afflictions or whatsoever arise from holding the view that the self is real. Those who practice dhyāna yoga do not consider the self as real, because of the wisdom gained from the yoga practice. They do move away from the view that the self is real and stop karmic actions.

It is essential to understand that *mirages* and *dreams* are all *actual* phenomena, which actually happen and have actual consequences. However, that does not mean that they happen to us in a nondeceptive way [48]. The road mirage (inferior mirage) gives an illusion of the reflection of a blue sky on the heated distant ground, but it is a real *optical* phenomenon. This is really the bright sky seen in the distance, which appears to be reflected in a water puddle acting as a mirror. In reality, the road mirage is not a reflection from the surface of water puddle, but refraction of the image rays through the heated air layer of changing density. There is no question of whether the dream is real or not. In fact, it is an *actual* physiological phenomenon occurring in our brain, but what the dreamer perceives is an illusionary view without any inherent nature.

By analogy, karmic action and its consequence are real empirical phenomena like mirages but are empty in nature. Because they lack inherent nature, they neither are permanent existences nor become non-existence (because they were not really existent since the beginning). Hence, with respect to mirages or dreams, one cannot have a view of permanent existence or a view of annihilation. Please see the description from the *Ratnakuta* Sutra in Section 9.5 (5).

It would be interesting to compare Buddhists' aforementioned view on the karmic actions with Mary Shelley's *Frankenstein*. If Mary Shelley is compared with the man of magical power, the illusory body created by the person may correspond to Frankenstein. The living monster created out of exhumed corpses may correspond to the secondly illusory body created by the first illusory body. However, there exists an essential difference between them. Buddhists' view applies to everything in this world that lacks intrinsic nature, whereas Frankenstein's story is an imaginary horror story. From Buddhists' view, the same applies not only to a horror story but also to a comedy or a love story. The latter are all actual phenomena, which actually appear and have actual consequences in the samsāra, the *mundane* world.

11.6 What is Modern Science — 2

1900s was Einstein's century

In 1900 at the turn to the 20th century, Lord Kelvin, a respected scientist, stated, *"There is nothing to be discovered in physics now. All that remains is more and more precise measurement."* However, five years later, the year 1905 is often called Einstein's "wonderous year." Albert Einstein (1879–1955) published four groundbreaking papers within a year. Each of them was a revolutionary work that contributed substantially to the foundation of modern physics and changed our views on space, time, mass, and energy: (i) *Brownian motion*, which settled the debate over the existence of atoms and laid the foundation for a new field of work known as statistical mechanics; (ii) *photoelectric effect*, which demonstrated the particle aspects of light and led to the quantum theory of matter; (iii) *special relativity*, which overturned a model of space and time that had stood for millennia; (iv) *mass–energy equivalence*, which connected matter and energy and led us to a true understanding of the huge amount of energy released by stars.

The third paper of relativity reconciles Maxwell's equations for electricity and magnetism with the laws of mechanics by introducing major changes to mechanics close to the speed of light, c. Later, this was known as Einstein's special theory of relativity. Einstein puts forward two postulates to explain experimental observations. *First*, he proposes the principle of relativity, which states that the laws of physics remain the same for any non-accelerating frame of reference (an *inertial frame*) and applies it to the laws of electrodynamics as well as mechanics. *Second*, Einstein proposes that the speed of light has the same value in all inertial frames of reference, independent of the motion of the emitting body. In Michelson–Morley's experiment performed earlier in 1887, they could not discover any motion of the Earth relative to the "light medium." The speed of light, c, is fixed and thus not relative to the movement of the observer. This was impossible under Newtonian classical mechanics. The theory suggests that the phenomena of electrodynamics and mechanics possess no properties corresponding to the idea of *absolute rest*.

In the fourth paper, Einstein considered an equivalence equation stating that a particle of mass m possesses energy E (the *rest energy*) and deduced what is arguably the most famous of all equations: $E = mc^2$, where c is the speed of light. This means that if a body gives off energy ΔE in the form of radiation, its mass diminishes by $\Delta E/c^2$.

This was a glorious triumph of the physical theory and also opened the atomic age.

In the Newtonian theory, the gravitational force at a given point and given time depends on the current positions of all masses (at the same time). Therefore, a sudden change in any of these positions produces an instantaneous change in gravitational forces everywhere, which means that the change of the gravitational field propagates at infinite speed. This is in conflict with the principle of Einstein's 1905 *special theory of relativity*, that no influence can travel faster than light. According to Einstein's general theory of relativity proposed in 1915, a sudden change in the position of a mass will produce a change in the gravitational field which propagates at the speed of light, c, to greater distances. Newton's notion of absolute space and time has been replaced by the *general theory of relativity*. The theory is based on the gauge principle and claims that the underlying laws of physics remain unchanged in all reference frames, irrespective of their acceleration or rotation.

Based on the mass–energy principle $E = mc^2$, top scientists developed atomic bombs during wartime in 1945. As a matter of fact, two atomic bombs were dropped over Hiroshima and Nagasaki in 1945. Evidently, Einstein, the gifted physicist and innate pacifist, was distressed and deeply hurt by this. Earlier, during the development stage of the Manhattan Project, he was not invited to the project, because he was an immigrant to the United States from Germany. He settled in the United States and become an American citizen in 1940. On May 8, 1945, World War II ended in Europe when the Nazi regime surrendered unconditionally. Then, Einstein opposed the use of atomic bombs although he had written a letter to the US president earlier suggesting the possibility of the creation of an atomic bomb. Anyway, atomic bombs have *completely* and *revolutionarily* changed the world of postwar generations.

Einstein addressed the Caltech student-body in 1930 on how science had not yet been harnessed to do better than harm [89]:

During the war, it gave people "the means to poison and mutilate one another," and in peacetime it "has made our lives hurried and uncertain." Instead of being a liberating force, "it has enslaved men to machines" by making them work "long wearisome hours mostly without joy in their labor." Concern for making life better for ordinary humans must be the chief object of science. "Never forget this when you are pondering over your diagrams and equations!"

11.7 Natural World, Dependent Origination and Śūnyatā Nature

Dependent origination

Every phenomenon in nature around us is characterized by *śūnyatā*. Before presenting various phenomena in the natural world, we cite how the dependent origination is interpreted by a Buddhist Nāgārjuna (Section 9.3). The dependent origination is equivalent to the *śūnyatā* nature. He explained the concept of dependent origination as *"whatever arises in dependence on something else does not arise as inherent existence,"* and interpreted the dependent origination in an insightful manner with the following *eightfold negations* (i.e. negated with four pairs of representative expressions):

> *Whatever is dependently arisen is*
> *Neither ceasing to exist, nor arising. Neither terminable, nor permanent,*
> *Neither self-identical, nor distinctive. Neither coming, nor going,*
> *And free from conceptual fabrication.*

According to Buddha, *"Whoever sees dependent origination sees the dharma; whoever sees the dharma sees dependent origination."*

Let us see a biological example first, an interesting phenomenon of sprouting from a plant seed, and then consider some other examples from physics.

(a) *Sprouting of a plant seed*

Let us observe the amazing nature of seed-sprouting. Consider the germination sequence of a plant seed shown by the picture Fig. 11.1, from the point of view of *dependent origination*. Seed germination depends on both internal and external conditions. Important external factors are *water, oxygen,* right *temperature*, and *light*. The germination starts within a seed; it results in the formation of the seedling. A pair of new leaves (in dicots) sprouts from a seed.

From the philosophy of *dependent origination*, the pair of leaves has just sprouted up within the seed and not come from outside. The seed did

Fig. 11.1   A sequence of plant germination from a seed [99].

not go anywhere but just transformed to the seedling, i.e. sprouted. The new leaves have not arisen as self-existence but soon transform to other forms in due course. This process is not an intrinsic self-existence forever. Obviously, the dicot-state is not the same as the original seed, but it is not completely different from the original seed, because it does never come up (say) from an egg. Without the original seed, the dicot shown in Fig. 10.1 never comes up. It is obvious that the seed is neither permanent nor terminable since its germination and subsequent growth are an endless sequence of plant processes.

This observation is consistent with the *eight negations* given in the beginning. The sprouting and subsequent growth is as follows:

> *The plant germination is neither ceasing, nor arising from empty space,*
> *neither terminable nor permanent.*
> *Its state is neither self-identical, nor distinctive. neither coming from*
> *outside, nor going,*

The sprout, which is transformed from a seed, is not the same as the original seed itself. The plant object is neither perishable (it does not stop being) nor permanent (it does not keep the same). The sprout arising from a seed is not generated from itself (not identical to the seed); nor is it wholly other than the seed; it does not arise spontaneously from empty space; it is not created by God. It is said:

> *Whatever arises (comes into existence) dependently on something else*
> *(not arisen by God)*

cannot be an inherent existence. It neither stays to exist permanently, nor ceases to exist completely.

Namely, the eight negations are not the nihilism at all, but they describe real active phenomena of nature, which is beyond the power of analytic wording. Only negated expressions can do it, or photographic pictures can do it.

It should be noted that the aforementioned observation of sprouting and subsequent growth is accepted by both the Buddhistic view and the scientific view.

(b) *Life of a tropical cyclone (typhoon or hurricane)*

Typhoons or hurricanes are huge atmospheric eddies, usually born in tropical areas and decay in higher latitudes (Fig. 11.2). They are born autonomously within the Earth's atmosphere and end their whole lives within a month or so, diffusing into the same Earth's atmosphere. Modern computers can simulate the whole processes from cyclone genesis to their final diffusion and extinction.

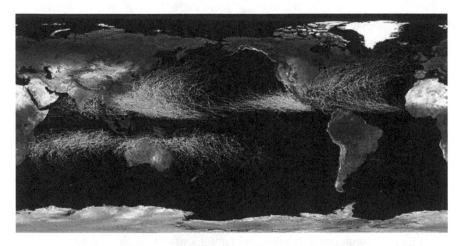

Fig. 11.2 Global tropical cyclone tracks between 1985 and 2005 [90], indicating the areas where tropical cyclones are born usually in tropical areas. They migrate to the east first and then away from the tropical region, guided by winds, and finally diffuse in higher latitudes [89].

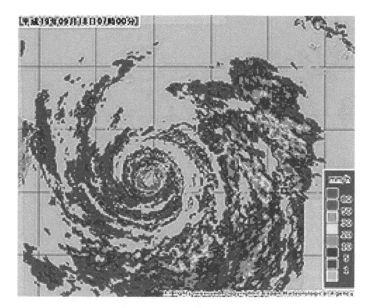

Fig. 11.3 Spiral bands of raining [93].

They are dependently arisen within the atmosphere. They neither come from outside nor go anywhere else and thus never exit the atmosphere. As interpreted in the following text, the whole life of a tropical cyclone is a continuous sequence of dependent originations (hence phenomena of emptiness).

Let us consider *cyclones,* important and influential objects of meteorology. What are the tropical cyclones such as typhoons or hurricanes? They are large-scale atmospheric eddies like the one shown in Fig. 11.3, which bring strong wind and heavy rain around their central core, called *eye.* A *typhoon* is a storm that occurs in the northwestern Pacific Ocean, a *hurricane* occurs in the Atlantic Ocean and northeastern Pacific Ocean, and a *tropical cyclone* occurs in the Indian Ocean or South Pacific (see Fig. 11.2). They are given different names depending on the locations of their formation, but they are the same meteorological objects, a *tropical cyclone* (TC) [89, 90].

In Fig. 11.2, one may see that tropical cyclones emerge from a *non-cyclone state* in the tropical area and diffuse into another non-cyclone

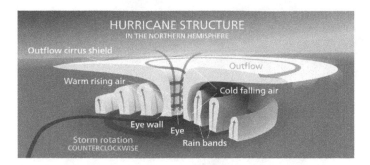

Fig. 11.4   Internal structure [90].

state at higher latitudes. Figure 11.2 also shows the tracks (loci) of the core eyes of a number of tropical cyclones (TCs). There exists a clear belt along the equator where no loci of TCs are observed. This is because the Earth's rotation around the north-south axis is not effective there for the formation of TCs geodynamically. But the formation of TCs is most effective at some (north or south) latitudes lower than 30° (away from the equator).

In meteorology, a hurricane (a tropical cyclone) is in general a large-scale air mass that rotates around a central air pillar of low atmospheric pressure. Figure 11.3 exhibits the RADAR image of spiral rain bands in a typhoon. The inward spiraling wind bands rotate about the clear quiet air pillar of low pressure, observed as an eye. Its internal structure is shown in Fig. 11.4.

There are tropical zones of trade winds on both sides of the Earth's equator, where tropical cyclones are formed (Fig. 11.2). The trade winds are easterly winds and blow predominantly from the northeast in the Northern Hemisphere and from the southeast in the Southern Hemisphere. As the air masses blow across tropical regions, they get moisturized by the vapor from the sea surface heated by strong sunlight. The heated moist air mass becomes lighter, and a cluster of clouds is formed over the heated sea, which is typically composed of cumulus clouds extending vertically as high as 4 km. In fact, the clouds and the surrounding air masses derive their heat energy through the evaporation process from the ocean. When moist air rises up, cooling down to the saturation level, and ultimately re-condenses into water droplets composing clouds, the latent heat of vapor that is acquired at the time of evaporation is released into the air around

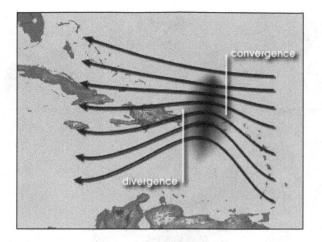

Fig. 11.5 Waves in the trade winds: Areas of converging winds create wavy pattern in the atmosphere that may lead to the formation of hurricanes [89].

the clouds. Thus, the air mass is warmed. It becomes lighter and rises upward even higher. The clusters of cumulus clouds are considered to be the seeds that have the potential to develop into a tropical cyclone when conditions are favorable.

The seeds (favorable conditions) for hurricane development are identified as wavy deformations formed in the zone of trade winds (Fig. 11.5). A hurricane emerges spontaneously in the otherwise quiet horizontal layer of trade winds (nearly parallel with continuous change vertically) from local wavy deformations over the tropical ocean. It is generated by a sequence of meteorological mechanisms composed of several physical factors. Here it is emphasized that the hurricane is formed within the air mass of high humidity of trade winds, influenced by the Earth's rotation. A *cyclone* refers to the wind swirling around its central core. The influence of the Earth's rotation is obvious since their winds blow counter-clockwise in the *Northern Hemisphere* and clockwise in the *Southern Hemisphere*. In addition, a sufficiently warm condition is required (sea surface temperature exceeding 26°C) to fuel the cyclone system.

The sprouting of a seed for germination requires *water, oxygen, right temperature*, and *sunlight*. Just like that, a hurricane requires high humidity and sufficiently warm sea surface. In addition, to develop sufficiently, atmospheric deformation waves must accompany converging streams to a

Fig. 11.6 A sequence of hurricanes showing three growth stages of formation (right), infant (middle), and mature (left) [91]. Courtesy of NOAA, Live Science by Brett Israel, 2010.

focal core, and they must last for a sufficient period such that the stream pattern can feel the Earth's rotation, specifically for a period of sufficiently longer than 24 hours.

Figure 11.6 exhibits a sequence of three hurricanes in their developing stages (three swirling cloud patterns), observed in the Atlantic Ocean's tropical area. The following is a citation from an article by Brett Israel (2010) [91]:

> *Many of the storms start as seedlings coming off the Africa. Low-pressure systems move westward off the west coast of Africa, where they encounter the warm waters of the tropical Atlantic, which serve to drive the convective engines that power these tropical storms. The warm waters of hurricane alley can sustain a storm all the way from its seedling stage in Africa to mature hurricanes. According to scientists, the ocean waters must be at least 26.5°C down to a depth of at least 50 m below the surface to properly spawn a storm.*
>
> *If conditions in the atmosphere are favorable and the waters are warm enough, a tropical depression develops into a tropical storm, then*

Fig. 11.7   Decaying stage of a typhoon [92].

finally forms a hurricane, which is not unlike a giant swirling mass of
thunderstorms. Hurricanes always rotate counterclockwise in the
Northern Hemisphere.

Upon landfall, the sustained winds can drop rapidly due to a dampening effect caused by rough terrain — mountains, hills, trees, and buildings. Or staying over the ocean at higher latitudes, the cyclone has its energy supply reduced by running over colder water, when it reaches the decaying stage (Fig. 11.7). It moves into a region of increasing wind shear of the westerly at middle latitudes, which disturbs the vertical structure of the system leading to the disintegration of the cyclone system.

Thus, a tropical cyclone originates dependently in an otherwise quiet tropical area from wavy disturbances over a sufficiently warm sea surface supplying abundant moisture, which continue for more than 24 hours, and diffuses into the atmosphere at much lower temperatures. Cyclones emerge from the geo-atmosphere, experience continuous transformation during growing periods, and finally decay into the geo-atmosphere. Thus, the lifetime of a cyclone is said to be continuous *dynamical transformation of emptiness*. There is nothing that exists permanently.

11.8 Scientists and Buddhists

One of the targets of study in modern physics is the natural world and its structures, which are investigated by the language of *mathematics*. The evolution of a physical system in nature under consideration is described with the principles of physics and mathematics. In a physical study, first, a particular system is chosen or defined. Then under well-defined conditions, the system is investigated in detail both experimentally and mathematically, so that the state evolution is captured with less ambiguity.

It must be noted that for the success of a scientific study, a particular well-defined system must be chosen. However, everyday life is not always well conditioned; sometimes it proceeds ambiguously and sometimes it does not proceed as planned. This means that everyday life is often out of the scope of a scientific study and slips out of science because it is not a well-defined phenomenon.

When we carry out a scientific activity such as the observation or analysis of natural objects, there is an independent subject (person) who observes it or analyzes it. That is a self, believed to be a solid existence by scientists and their societies. The self does a scientific work and accomplishes a scientific contribution. Hence, a scientific activity is not separable from the activity of a subject-self, although nature as an object shares the property of *dependent origination* and is regarded as a *śūnyatā* object. This is seen in the examples presented in Section 11.7.

On the other hand, Buddhists practice the *six pāramitā*s (Section 8.6), which are inseparable from *śūnyatā* (emptiness) and *anātman* (no soul). Among the six *pāramitā*'s (listed in Section 8.6), the first four may be carried out by good scientists, which are (i) generosity, (ii) noble conduct, (iii) patience, and (iv) steadfast effort. However, the remaining two, (v) meditation (*dhyāna*) and (vi) *prajñā*, may be the practice of Buddhists, which are not required for good scientists.

Scientists' activity is the study of nature by means of observation, experiment, and analysis to find scientific conclusions. On the other hand, Buddhists study not only the surrounding outside world but also their internal one (mind). Regarding the perspectives toward the surrounding world, Buddhists mostly share a common view with scientists. However,

Buddhists consider their inside world more significantly than scientists do.

In fact, Buddhists have a wider perspective internally on this world, and follow the way of *prajñā-pāramitā* (practice of wisdom of transcendence). This practice emancipates one from worldly attachments by *śūnyatā* action and transcends any conceptual fabrication. Buddhists wish to liberate their minds and to have peace and wish the same for all living beings under *mahāmaitri* and *mahākaruṇa*.

Part III

Bodhidharma in Kancheepuram: Commemoration

Bodhidharma in stone relief carving
at Vaikuntha Perumal Temple, Kanchipuram, Tamil Nadu, India.

Chapter 12

Bodhidharma from Kanchipuram

12.1 How the Personality of Bodhidharma and his Kingdom was Described in China

According to Chinese literature, when Bodhidharma was staying in Shaolin Monastery in the 530s CE, his disciple described him as follows:

> The Dharma Master Bodhidharma was from the South India. He was the third son of a great Indian King. He was extraordinarily wise and bright as a child. He embraced the Way of Mahayana.

The name of his home kingdom was not given, except for "from South India." However, in modern literature, the name of his kingdom in South India is written with three Chinese characters, 香至国. But, so far, almost nobody has definitely described its detailed location in India and the corresponding Indian name. How that happened is, in fact, a riddle. Having already renounced secular life, the monk Bodhidharma might not have been interested in telling *which Indian kingdom he was from.*

Based on detailed investigation of Chinese literature and recent information on the internet, the present author thinks that the actual story might have been as follows.

On the China side, the oldest literature that had the name 香至国 appeared more than 400 years after the time of Bodhidharma. That book was *Record of Patriarchal Hall* (祖堂集) [55], published in 952 CE, and the next ones were *Record of Transmission of Dharma Lamp* (景德伝燈録) (1004 CE) [54] and *Record of Transmission of Orthodox Dharma* (伝法正宗記) (1062 CE) [60]. However, despite being lost during the long history, the first description might have been in the lost volume 7 of the book *Treasure Forest Record* (宝林伝) (801 CE) [59], as noted by the translator of the book (S. Yanagida) 祖堂集 [55] on page 91. Hence, it is likely that the name 香至国   was not known publicly in China before 800 CE.

With a dialect of Fujian Province (福建省, the Min-nan language), the name 香至国 is pronounced as "*Kang-Zhi-Guo.*" Hence, it is understood to refer to the ancient capital town "Kanchee-puram," where the third letter, "*Guo,*" means "country" (*puram* in Tamil). Thus the name 香至国 denotes nothing but phonetic representation in Chinese of the town name Kanchipuram.

It is interesting to see how Bodhidharma was depicted by two main disciples. The earliest description of Bodhidharma was given by the monk Tanlin (曇林 engaged in translation of Sanskrit texts) in 530s CE, who described his teacher as follows:

> *The Dharma Master Bodhidharma was from South India. He was the third son of a great Indian king. He was extraordinary wise and bright as a child. He embraced the Way of Mahayana.*

At nearly the same time, somewhat later than Tanlin, the monk Huike (慧可, Second Zu, who succeeded the Dharma of Bodhidharma) depicted his teacher:

> *Our Teacher had come from the West* (India) *to the covered in East* (*China*). *When eating, he was exposed to winds; when sleeping, covered in dewdrops. Never had recognized how many seasons passed for his travel. Sailing over the sea and walking in the mountains. Never had recognized how many storms attacked, and difficulties came over. How does such a person get exhausted?*

(See footnote in § 10.3.3)

The following pair of visual images in Fig. 12.2 are seen resonantly with the above two descriptions found in Chinese literature. The photo on the left is of a sculpture relief found in India at the Vaikuntha Perumal Temple (Kanchipuram), and the photo on the right is of a painting placed in the guest room of the Tian-Tong Temple (天童寺, China), which the author visited in 2009 during his own pilgrimage as a Zen Buddhist and

(a)   (b)

Fig.12.1   (a) Bodhidharma in relief (2019) at Vaikuntha Perumal Temple, Kanchipuram, Tamil Nadu, India. (b) Bodhidharma in a painting (2009) at Tian-Tong Temple (天童寺, China). Photographs taken by the author.

took the photo while waiting the Master.[1] Both of these show Bodhidharma in the form of a traveling monk. In particular, one can observe a certain similarity between the two images concerning the monk's style. It is surprising to consider the locations of the two temples where these were found. The temples are separated by a large distance, with the left one in India and the right one in China. The present author is very much interested in the similarity and wonders what kind of historical dramas were there in the background of these images.

---

[1] A young Japanese Buddhist monk, Do-Gen (道元), had visited this Tian-Tong Temple (天童寺), south of Shang-hai, and met the Chan Master, Ru-Jing (如浄禅師). Believing that he was a true master, the monk Do-Gen asked Master Ru-Jing various questions in order to ensure his understanding about Buddhism. The monk Do-Gen had succeeded the Dharma of the twenty-third Zen Master Ru-Jing and become the twenty-fourth Zen Master Do-Gen (道元禅師), who founded the Japan Soto School (日本曹洞宗), after he returned to Japan in 1227. The present author (T. J. Kambe) was ordained as an *upāsaka* of the Soto Zen School in 1966.

Fig. 12.2   Extended panel consisting of five different scenes (see text below).

The left image (a) in Fig. 12.1 was found as a part of a larger panel (Fig. 12.2), consisting of five different scenes (see below), which was placed on a wall along a long corridor surrounding the structure of the Vaikuntha Perumal Temple. It is said that the temple was constructed in the middle of eighth century CE by the Pallava king Nandivarman II. The corridor walls exhibit a number of inscribed panels, as if proud of them, with sculptures portraying the lineage of the Pallava dynasty, the actual kings themselves, battle scenes, and other events during the dynasty's reign. The arrangement of the lineage was up to the accession of the founder of the throne, Pallavamalla (i.e. Nandivarman II).

Four of the five different scenes in Fig. 12.2 relate to royal life and related events. The fifth image of Bodhidharma stands in sharp contrast to the royal scene. On the extreme left, one sees the father, King Simhavarman II (the throne from 436 to 460), who was a patron of Buddhism. The central image on the top shows a war scene with elephants (upper left), horses (upper right), and soldiers. The top right image portrays two elder brothers of Bodhidharma. The bottom right image depicts battle practice. The fifth image, in the lower central part, is of the third son (Bodhidharma, c. 440–536 CE).

The Vaikuntha Perumal Temple was built by King Nandivarman II (*Pallavamalla*) (reign: 731–c. 796 CE) of the Pallava dynasty. This implies that the story of Bodhidharma's family had been kept as a memory in the Pallava dynasty for nearly three hundred years in Kanchipuram, India. King Nandivarman II was born into a local dynasty of Pallava origin in the oversea country Champa (modern-day Vietnam) and was elected as a Pallava king (named Pallavamalla) at the age of twelve. The suffix "-malla" denotes "powerful man" or "athlete." See "The Pallava King from Vietnam" on the recent website (2017) of Live History India [94].

12.2 Ancient Relationship Between Tamil, Southeast Asia, and China

From ancient times, India has had close connections with China and Southeast Asian counties. In the present context, Vietnam was known in India as Champa, which was located in the central and southern parts of the present-day Vietnam. Indian merchants regularly traded with ports not only at Champa (Vietnam) but also at Guangzhou (China, see Section 10.3.1) and Quanzhou (China, see below) and other ports in Southeast Asian counties. They even settled there. In fact, one of the oldest pieces of epigraphy found in Southeast Asia is the *Vo Cahn* inscription found in the village Vo Cahn in central Vietnam, and it dates back to the 7th century CE. This is a Sanskrit inscription written with the Pallava Grantha script used at the time of the Pallava dynasty from 6th to 9th century CE in South India. Hinduism adopted in the 4th century CE shaped the art and culture of the Champa for centuries. See [94] for the connection with Pallava kings.

The fact that King Nandivarman II, who built the Vaikuntha Perumal Temple, had come from Champa (modern-day Vietnam) gives a special meaning to the relief sculpture of Bodhidharma installed on its wall exhibiting the lineage of the Pallava dynasty. It would be worth investigating its historical and cultural backgrounds in regard to Bodhidharma.

Another recent news item (2019) about a Tamil stone inscription in the *SimpliCity News* [95] is helpful in considering its background connection. In the port town of Quanzhou (泉州), located along the southeast coast of China, a Tamil stone inscription was found. It throws light on the centuries-old cultural relationship that existed between ancient Tamil and China. In Quanzhou, there exists a renowned Buddhist temple, the Kaiyuan Temple (開元寺), so renamed by the contemporary Emperor in 738 CE (originally built in 685), and hence spotlighted in those times. Behind its main hall, the Mahavira Hall, there are some remains related to the Vishnu temple built in 1283 by the Tamil merchant community in Quanzhou, shedding light on ancient maritime relationship between China and the Tamil community.

One more historical evidence is the book *Record of Patriarchal Hall* (祖堂集) [55], published in 952 CE, which was the first to record the name 香至国 denoting the Kingdom of Bodhidharma. The two Buddhist authors who compiled the book are important in the present context. They were Chan Buddhists Jìng (靜) and Yún (筠), who stayed at the Kaiyuan Temple in Quanzhou. In other words, Bodhidharma's Buddhism was practiced at the Kaiyuan Temple in the 900s CE.

The transmission of Bodhidharma's Chan Buddhism from Huike (the Second Zu) to later generations was successful in the 500s and 600s CE. From the 600s to 700s, the Chan school developed nationwide during the Tang dynasty (618–907 CE) of China. This new-style Buddhism was recorded enthusiastically by a number of publications such as [62] (645 CE), [78] (689), [61] (713), [56] (716), and [57] (774). In particular, the earliest was the *Memoirs of Eminent Monks Continued* [62] published in 645 CE. The stone monument [78] erected in 689 CE commemoratively stated, "The dharma teacher Bodhidharma brought the dharma of Shakyamuni to the East land from India," and inscribed the initial succession of the dharma by the names of Chan masters.

Thus, Chan Buddhism became very popular in China by 731 CE, when the young boy Pallavamalla of Champa was crowned as King Nandivarman II (reign: 731–c. 796 CE) of the Pallava dynasty in Kanchipuram, India.

It is very likely that the Tamil traders in Quanzhou were informed of the name of the founder of Chan Buddhism, Bodhidharma, who came from South India. The Tamil merchants were probably then asked by their Chinese customers:

"*Do you know the Indian monk Bodhidharma who brought the Shakyamuni's dharma to China?*"

Perhaps, an Indian merchant replied:

"*He was a prince of a king of the Pallavas,*" whose capital was "Kanchee," phonetically written in Chinese letters as "香至".

Then, much later (952 CE), influenced with this kind of dialogue, Chinese Buddhists wrote:

第二十八祖菩提達摩和尚者　南天竺国香至大王第三子也 ([55] (c), p. 459)
　　The 28th Zu, Bodhidharma Osho, was the third son of the Kanchee King of South India.

In other words, the father king is named as "Kanchee King of South India" (香至大王). However, the first description might have been much earlier in 801 CE in the lost volume 7 of [59], as remarked in Section 12.1.

The following story line is conceivable now. King Nandivarman II, who constructed the Vaikuntha Perumal Temple, was informed by Tamil traders about a historical episode in China — that a new school of Chan Buddhism was developing there and its founder was Bodhidharma, the third son of the Great King of South India. From the viewpoint of the present writer, this could be one of the good reasons for King Nandivarman II to commemorate Bodhidharma in his home kingdom.

Chapter 13

Commemoration Project

13.1 Brief History of the Project

A non-profit organization (NPO), World Association of Bodhidharma (WABD Japan), was established by Zen Buddhist monks and general Buddhists in Japan in 2009 with the aim of commemorating the great Buddhist monk Bodhidharma. The Tokyo Legal Affairs Bureau registered the NPO WABD Japan in 2010 as a legal body of a corporation engaged in non-profit commemoration activities. In 2002, a related body, "Japan Daruma Kai" (日本達磨会), had a cooperative project with a Chinese body of Buddhists to restore the Kong-xiang Temple (空相寺) at the foot of Bear-Ear Mountain in China. (See Section 10.3.8 and Fig. 10.7 of the photo of the Stupa of Bodhidharma).

In order to make a monument in India to commemorate Bodhidharma, the NPO WABD *Japan* had a contraction with the stone company GEM Granites, based in Chennai, India. In February 2014, a group of representatives of the Japan NPO visited *GEM Granites* for collaboration (Fig. 13.1) and then visited a quarry (in mountains of Tamil Nadu; Fig. 13.2) owned by the company. For the collaboration, the chairman, Mr. R. Veeramani, gave the quarry people detailed instructions on how to produce a single block weighing more than 20 tons in black granite, which is very rare.

Mr Veeramani helped the Buddhist team of NPO WABD Japan in holding a Buddhist ceremony to sanctify the quarry ground on February 26, 2014 (Fig. 13.3), before quarrying out Tamil stones to be used for the

Fig. 13.1   Front face of the Bodhidharma Memorial (2015). [Monolithic black *granite-stone* quarried from a Tamil mountain].

Fig. 13.2   At the Garden of GEM Granites. Mr. R. Veeramani, *Chairman*, at the center.

monument.[1] A single large monolithic black *granite-stone* was mined from this quarry and used as a raw stone for the commemoration monument.

In 2015, the stone was brought to the GEM factory in Chennai (Tamil Nadu), and the Memorial stele (Fig. 13.1) was fashioned from it.

---

[1] In fact, Mr. Veeramani's family loves Buddha and Buddhism and offers flowers daily.

Fig. 13.3 Buddhistic sanctifying ceremony at the quarry (in Tamil). Ven. Kikuchi is sitting at the center and Ven. Watanabe standing behind, on February 26, 2014.

Fig. 13.4 Quarry of GEM Granites.

Fig. 13.5　In front of the monument just made at the GEM factory: From the left, Dr. Ven. Jinwol Lee (Dongguk University, Korea), Kung-Fu Master Ven. Shi Yan-Lin (釈延琳, Shaolin), and T.J. Kambe. (August, 14, 2014.)

13.2　Bodhidharma Foundation and its Land

In order to place the Bodhidharma Memorial Stele in India, *his home*, it is necessary to establish a managing foundation in India and secure a land for the installation of the Memorial. On March 18, 2019, our foundation was approved by the Government of India under the name Bodhidharma Dojo Foundation (BDF).

It is a non-profit and non-governmental organization in India.[2] The GEM chairman, Mr. R. Veeramani, accepted to assume the presidency of BDF. A plot of land to be used for placing the monument was acquired in the Kanchipuram region of Tamil Nadu in India. The land was named Bodhidharma Memorial Park (abbreviated as BDMP). On March 6, 2019, an Indian-style *bhoomi pooja* service was held at the undeveloped land

---

[2]BDF: based in Tamil Nadu. *President*: R. Veeramani; *Chairman*: T. J. Kambe; and *Secretary*: Uma Balu.

Fig. 13.6 *Bhoomi pooja* on March 6, 2019. Second and third persons from the left are a Hindu monk (leading the pooja) and Ms. Uma Balu.

that had just been acquired (Fig. 13.6). This is a ritual dedicated to Mother Earth, performed to pour energy into the land and pave the way for a smooth land construction. About fifty people attended this festival, including Buddhist citizens, some local influential people, a real estate agent, and the previous owner. The monument itself was completed in 2015 and kept safe at the GEM factory in Chennai. Without delay, it was transferred to the Bodhidharma Memorial Park and temporarily installed. On March 26, 2019 (Fig. 13.7), a delegation from the Japanese NPO WABD Japan visited the site and held a Buddhist ritual for opening the *Memorial Monument* at the home land of Bodhidharma, according to the Zen Buddhist Way. Thus, the Memorial Monument (Fig. 13.8) was finally installed and fixed at the Bodhidharma Memorial Park on July 18, 2019. It was fixed on a solid base semipermanently.

The Bodhidharma Memorial Park (BDMP) is located in a quiet rural area in the suburbs about 6 km southwest from the town center of Kanchipuram City. The land of BDMP is flat and surrounded by

Fig. 13.7 *Ritual for Opening the* Monument by NPO WABD Japan with eight Venerables: *Ichikawa (President, front-center), Hozumi, Sano, Kikuchi, Watanabe, Ohno, Utsuno* and *Fukuoka, and four* secular Buddhists: *Hirai, Tanaka, Fukuda* and *Kambe*. On 26/03/2019.

Fig. 13.8   Memorial monument placed at the Bodhidharma Memorial Park.

Fig. 13.9 The Bodhidharma Memorial Stele was placed and fixed firmly at BDMP on July 18, 2019. Second from the left is Mr. Krishna Kumar (director of BDF).

agricultural land, covering an area of about 6000 (about 1.5 acres). The location address of Bodhidharma Memorial Park is as follows:

Bodhidharma Memorial Park
No. 34, Kolivakkam Village, Kanchipuram District, Tamil Nadu, India

This place is a suburb when viewed from the current city area, but culturally it is known as a historical area dating back to the time of Bodhidharma. Hence it is a land suitable for the commemoration site. It is fortunate that local officials (and real estate agents) were understanding of our NPO activity. The town of Kanchipuram is one of the sanctuaries of Hinduism.

The happy, friendly circumstances are thanks to the Great Dharmachaya Bodhidharma himself and his noble grace. On July 18, 2019, the Memorial Monument was placed and fixed firmly in its proper place in the Bodhidharma Memorial Park. In 2021, future plans of the site were considered by all members concerned for its further development.

Chapter 14

Bodhidharma's Messages to Disciples

14.1 Four-line Holy Verse and Authentic Texts

Bodhidharma's message is well known by the following verse (Section 10.3.8):

> *No dependence upon words and letters,* 不立文字
> *Special transmission outside the scriptures,* 教外別伝
> *Directly pointing to the mind dharma,* 直指人心
> *Having insight into suchness, Attaining Buddhahood.* 見性成佛.

This is called, in general community, as Bodhidharma's essential verse, or *four-line holy verse.* However, its original source or the name of person who composed it are not clearly known. It is, in fact, puzzling to the present author. Section 14.3 proposes its possible source and a possible Buddhist monk according to the author's point of view. It is seen that the possible source was the text 《洗髄経》 (*Wash-Marrow Classic*) and that the possible Buddhist was the Second Zu, Huike Dashi.

Bodhidharma's messages are mostly known through legends of impressive dialogues with disciples or lay Buddhists,[1] as well as how he practiced, as expressed by the "nine years of wall-gazing" in a cave at the Shaolin Monastery. However, very fortunately, it is known that three texts have been left to us that are believed to describe his authentic messages with written letters (therefore very, very precious and new to most of us). It is miraculous that we can learn the teachings of the Buddhist monk (Bodhidharma) in the 6th century through written texts, although they are written all in Chinese characters.

---

[1] Lay Buddhists such as Emperor Wu (Section 9.3.1) or the local governor Yang Xuan-Zhi (Section 9.3.8).

Here, we will look at two of them. The first (Section 14.3) is a text [75] kept at the Shaolin Monastery: 『翻訳《洗髓経》 *(Wash-Marrow Classic)*』.[2] It is the Chinese translation of Bodhidharma's original manuscript in Sanskrit by the Second Zu Huike Dashi. Very Precious!

The second is described in Section 14.5, its formal title being 『略弁大乗入道四行 *(Concise Guidance on Mahāyāna Entrance Ways and Four Practices)*』, commonly abbreviated as *Two Entrances and Four Practices* (二入四行論) [51].

Those texts were not written in an Indian language, but in Chinese. How could it happen? According to Section 10.3.3, Bodhidharma is thought to have arrived around 527 CE at Shaolin Monastery, located just south of the Yellow River in a central flat area of China, and passed away in 536 CE.

After spending nine years quietly and patiently at Shaolin, Bodhidharma was finally satisfied with the distinguished Chinese disciples he had acquired. Two prominent scholar monks stand out among them. Needless to say, the first was the monk Huike (慧可), Bodhidharma's most trusted disciple. Huike (487–593) is now regarded as the Second Zu of the Chan (Zen) Buddhism lineage. Another one was the monk Tanlin (曇林), who was not only a disciple of Bodhidharma but also a translator and Sanskrit scholar. In fact, Tanlin was critical in solving the above riddle. The monk Huike, too, a close but elder colleague of the monk Tanlin, was also learned in Sanskrit.

14.2 How Bodhidharma was Described by his Disciples

The first disciple, Huike (Second Zu), wrote about his teacher as follows [75]:

Our Teacher had come from the West (India) to the East (China). During his meals, he was exposed to winds; during his sleep, he was exposed to

---

[2]Another one was the *Change Muscle-and-Tendon Classic* (*Yì-jīn-jīng,* 易筋経). See Section 10.4.2.

dewdrops. Sailing across the sea and walking in the mountains, He never realized how many seasons passed. He never realized how many storms attacked him and how many difficulties had arisen. How can such a person get exhausted?...

In this way, Huike Dashi (Great teacher) described, in the Introduction to the text《洗髓》, about his teacher, and added further *what Bodhidharma had in his mind as*: If one could find in China such a person who is receptive to the Supreme Way, following on that Way subsequently, the person must be suitable enough for entrusting the noble transmission. This is *the essence of the Noble Object that the Great Teacher had come from the West* (see Section 10.2).

First he came to Shaanxi Province 陝西省 and to Dūnhuáng 敦煌 (nearly 2000 km apart), leaving bowls (staying) at temples. Then, he came to Shaolin in the Central Plain (the flat land of the middle reaches of the Yellow River, centered on the region of the capital, Luoyang), where he spent nine years sitting cross-legged facing the wall . This was certainly a Buddhist practice of calming the mind and awakening. On the other hand, he had traveled all the way to China with the noble aim of transmitting Shakyamuni's Mind Seal to disciples, which he had not yet achieved. Hence, the practice was not a simple exercise of calming the mind and awakening, nor was it a simple meditation. For the time being, he was sitting quietly and keeping long in that position. In this way, he had been waiting for the visit of wisemen seeking the Way.

What the Great Teacher showed to people was the First Principle. Many listeners stuck to their own (secondary) opinions and traditions. They could not realize the true meaning and did not ask for further instruction.

I wondered who this monk was. I was moved by the impression that this person might be showing a profound Way. Fortunately, I was able to serve near the monk (around 528 CE), listen to his talks, and sincerely served him, asking about his teaching. Thus, I spontaneously (unintentionally) attained the sense of *Unsurpassed, Authentic Awakening*.

We found two volumes (cloth-bound books):《易筋》(*Yì-jīn, Change Muscle-and-Tendon Classic*) and《洗髓》(*Xǐ suǐ, Wash-Marrow Classic*). the meaning of the latter《洗髓》is deep, detailed, and advanced. It is difficult for beginners to understand it without first learning the fundamentals. Its effective result is difficult to achieve. It is the ultimate to be gained in the end.

Once achieved, it is well hidden and well appearing, stringing gold and penetrating stones, piercing the marrow and passing through it smoothly, unfolding the unbounded emptiness of the soul, collecting things and forming a shape, dispersing things then forming wind, but it cannot be grasped overnight.

The text 《易筋》（*Yi-jīn, Change Muscle-and-Tendon Classic*） is simple in its meaning. Get started, and get clues. It's easily understood by beginners. Its effect is easily achieved. It is regarded as the first to gain the base, which must be the fruit result of "*Yi Jin*." Specifically, by means of it, you must wash the marrow.

Thanks to the advice of our Teacher, it is already effective by practicing the *Yi Jin*. The original text of *Yi Jin* is stored in the wall of the Shaolin room, waiting for an interested person to find and obtain it. However, regarding the text Xǐ suǐ 《洗髓》(*Wash-Marrow Classic*), I am bringing it with me, attached to clothing (robe) and bowl, for my travel as a Buddhist walking like cloud-water (雲水)! After the practice was completed, what was acquired as a fruitful achievement was a great effect that had not been imagined before. However, I have not told it lightly to others. But I was afraid; if it was lost after, it would result in failing to meet the aim and expectation of the Teacher who had come from the West. Therefore, I wished to translate the text. But to make the ambitious attempt, I did not mind caring about my weak position and poor accomplishment that might go against the scriptures and dare not embellish the words. The preface is placed first in the front, and the main text is translated next in detail and placed backward. Thus, I wait for wise people to taste its deep meaning and gain something useful.

Inscribed Respectfully by Shi Huike

釋慧可謹序 [75]

14.3 Bodhidharma's Authentic Text I: 洗髓経

翻訳《洗髓経》(*Wash-Marrow Classic*) [75]

This is an English translation of a partial excerpt from the full Chinese text of《洗髓経》that Huike Dashi translated from the original Sanskrit text left by the Great Teacher Bodhidharma in his residential room at Shaolin. Much later than the times of both Great Teachers (Bodhidharma and Huike), Huike's Chinese version was kept at the Shaolin Monastery for internal use only. Fortunately, we can see it now. The *Wash-Marrow Classic* served as the original source for the historical development of Chinese Chan Buddhism. The Great Teacher Bodhidharma also left a companion text, *Yi-jīn* 易筋経 (*Change Muscle-Tendon Classic*), not given here.

Overview

Thus I have heard, Buddha told Subhuti: This stage can only be practiced after Yi Jin Gong (Yi Jin merit) is achieved. It is called the *quiet-night* session. It does not interfere with human affairs. During the day, monks are restless and busy tending to their clothing and eating. After the three meals are finished and the monk's work with water and fire has been completed, the evening stars are visible at dusk and dark rooms are lighted by lamps. After finishing their evening sessions, monks rest on beds, sleeping and snoring, forgetting living and dying. A wise one awakens alone and cultivates oneself quietly at night. Morning and evening, caressing the body and sighing, one day only goes by. Impermanence is coming quickly. A human being is like a fish in a small body of water.

Saintly and Ordinary the Same

Ordinary people pretend to be plausible, dress beautifully for body decoration, serve others' views, and eat dainty food day after day. People are all like this. In terms of their own affairs, mediocre. They do not mind life and death, led by fame and fortune.

Once fortune is gone, as if the oil had run out and the lamp had gone out, the corpse was buried in a wild field. The soul occupies a dream. Despite ten thousand sufferings and a thousand hardships, illusions never stop.

The sage is satisfied with the Truth alone, just wearing clothes and eating vegetables, not being greedy in order to stay healthy. Why is it necessary to become tired in order to satisfy the body and the mouth?

Penetrating into the sky and the earth, oneself is originally one with those. Although there is a difference in size between small and large bodies, their spiritual energy is the same. Heavenly cosmos has the sun and the moon, and the human body has eyes. The sun and the moon have bright and dark phases. Stars change like lamps. Even starlight varies in brightness.

On-going experience (*Kensho*) is never-ending and inexhaustible. Even a blind person reaches out their hand to touch the nose. The body is all eyes. Touching things enables us to know things to lean on. This is the essence of the alaya-mind, which includes heaven and earth. One can see without the eye and hear without the ear.

If the mind and heart are purified, they are not compelled by lust. Know where you are from, and return to where you originally were.

Ordinary people and saints are the same. Their eyes are horizontal and their noses long and straight. Their coming is the same, but their returning is different, because the ordinary ones are more interested in external affairs. If you can take care of yourself, think about life and death.

Taking advantage of this *washing-marrow* exercise, diligently practice the effort. **"Washing the marrow" makes us restore the original. The ordinary and the sacred merge into one.**

Walking, Living, Sitting, and Lying

Walk like blind person without a cane. Observing internally, intrinsic nature is illuminated. Foot lifting should be low and slow. Stepping firmly furthers progress. Every step should be all like this. Always be careful not to step too fast.

The world is busy making mistakes. Slow steps maintain peace of mind. Stay like a horse facing a cliff, or like a boat approaching a shore. Looking back, think. Looking down at your feet, recognize the position where you stand.

Do not separate from your present state of mind. Keep your mind off external affairs. Calming your body, you should know stillness and rest. Keep in mind and guard the emptiness. Keeping still, do not lean to one side. Straighten and fix your body. The mind, like still water or a mirror, calms the eyes and ears.

Things are entangled and moving in succession. **This very present moment is always the ultimate**. Monks sit like mountains overlapping. Standing straight is a solemn postures.

Close the mouth and keep the tongue deep inside. Breathe in and out of the nose. By repeating breathing, one returns to the greatness of one's origin like the sea. One's energy is activated with spirit, making it more open to itself and penetrating into the bones and the marrow. Thus immerse in meditation.

Sit with the legs crossed left or right, as appropriate. Both legs are always unevenly hooked, somewhat like a saw. Both hands should always be on the belly. Touching the navel, stroke the lower body. Caress the testicles like a dragon playing with a pearl.

When tired, sleep on your side. During sleep, do not be obsessed. Waking up, stretch your feet. There is no problem facing up.

Waking up from a dream makes no difference at all. Nine-year strenuous efforts. Beyond the world of life and death. Having realized the extreme essence of the Tathgata Mind.

Thus ends the chapter "Walking, Living, Sitting, and Lying." This is the very Truth.

Postscript (by Great Teacher Huike)

First, the main sutra was translated. Next, what was meant was presented. The descriptions in both cases are different, but their meanings are the same. "Dharma" is a Sanskrit word and "法空" is a Chinese term. All existing things are empty — that is, *non-touching* and *non-separating* simultaneously.

If one sticks to the sutra, one will never be able to get it. Things are different variously. Their change takes different trends and inclinations.

If something is the same as oneself, then approve it. If something is different from oneself, then destroy it. Only the aging and the dead are interested in the teachings of mud-religion. They have a distorted, contemptible mind. Sitting in a narrow well, they seek the sky.

Nothing hinders the Great Teacher, living freely as a bodhisattva. He traveled to the East and returned to the West, carrying only a single item. Nirvāṇa remains at Mt. Bear Ear. This is neither empty dust nor empty theory. Once great freedom is acquired, nothing hinders, nothing hooks.

Ah! My Great Teacher has known everything since birth. Since birth, he has understood everything without words. From a young age, he excelled extraordinarily.

When young, he traveled all over India and mastered various teachings and theories. **Without sticking to words (being only a device), he addressed the tenet essence.**

Having come to the East land. **Directly pointing the essential nature. Freeing entanglement and freeing binding**. Teacher from heaven.

Gratitude. The Teacher's grand mercy. Leaving the exquisite truth. Persons of a later generation, please never ignore this.

14.4 Source of "No Dependence Upon Words and Letters,...."

In the previous section of the text *Wash-Marrow Classic* (洗髓経), one can find the descriptions related to the *four-line holy verse*: "*No dependence upon words and letters, Special transmission outside the scriptures, Directly pointing to the mind dharma, Having insight into suchness, Attaining Buddhahood.*

- In the last section, **Postscript**, in the third paragraph from the bottom, there is the sentence "**Without sticking to words, he addressed the tenet essence**" (in boldface). This corresponds to "*No dependence upon words and letters.*"
- In the **Postscript** section again, in the second paragraph from the beginning, there is the sentence "**If one sticks to the sutra, one will never be able to get it**." This corresponds to "*Special transmission outside the scriptures.*"
- In the same section, in the second paragraph from the bottom, there is the fragment "**Directly pointing the essential nature. Freeing entanglement and freeing binding**." This corresponds to "*Directly pointing to the mind dharma.*"
- The last verse, "*Having insight into suchness, Attaining Buddhahood,*" is "見性成佛" in Chinese letters.

  The first part, "*Having insight into suchness,*" is expressed by 見性, which is **Kensho** in Japanese or **Jiàn xìng** in Chinese. This is described in the section **Saintly and Ordinary the Same**. In its middle part, **"Kensho"** is interpreted in boldface as follows:

Kensho is never-ending and inexhaustible. Even a blind person reaches out their hand to touch the nose. The body is all eyes. This is the essence of the alaya-mind, which includes heaven and earth. One can see without the eye and hear without the ear.

Thus, the **"Kensho"** is interpreted concisely by *"One can see without the eye and hear without the ear."* This corresponds to *"Having insight into suchness."*

The second part, *"Attaining Buddhahood,"* is found in the section **Walking, Living, Sitting, and Lying**. Seeing the part of boldface,

This very present moment is always the ultimate." And
Waking up from a dream makes no difference at all. Nine-year strenuous efforts. Beyond the world of life and death. Having realized the extreme essence of the Tathāgata Mind."

This addresses *"Attaining Buddhahood."*

Thus, it is seen that the *four-line holy verse* given at the beginning of this chapter has been interpreted by the text 《洗髓経》 *Wash-Marrow Classic*. In addition, the Buddhist who composed the *four-line holy verse* was none other than the Second Zu, Huike Dashi, since the Dashi was the person who translated the Great-Teacher Bodhidharma's original Sanskrit text into Chinese.

This finding seems to imply another interpretation of the 4th verse (see the last part of Section 10.3.8).

14.5 Bodhidharma's Authentic Teaching II: 二入四行論 [51]

略弁大乗入道四行 *(Concise Guidance on Mahāyāna Entrance Ways and Four Practices)* commonly abbreviated as *Two Entrances and Four Practices* (二入四行論)

(a) *Introduction by the disciple Tanlin*

Our Dharma Teacher had come from a western country, i.e. South Hinduka (South India, where Hinduka was derived from the Indus River). He was the third son of a great king. When a child, he was extraordinarily wise and bright, and clearly understood everything he heard. He embraced the Way of Mahāyāna and thus renounced the secular life to become a monk. Enhancing his spiritual life, he calmed the mind into *nirvāṇa*, tranquility. Observing worldly affairs with insight, he clearly understood the world both internally and externally. His virtue prominently surpassed that of people in society. Feeling deep regret for the weakness of the Authentic Teaching in remote places, he decided to cross oceans and mountains, and ventured to itinerate to the land of Han-Wei (漢魏, i.e. China). In the country, there exist wise people who do not necessarily bother about their own past ideas and do not stick to personal views. Among them, there is no one who does not believe in Bodhidharma. On the other hand, there are those who are trapped by external appearance and keep their own views. They criticize him and blame him.

Fortunately, there were DaoYu (道育) and Huike (慧可). The two *sramana*s (monks) had highly noble spirit despite their youth. They were very pleased to have met Bodhidharma as their Dharma Teacher. They practiced under his guidance for several years and asked respectfully for instruction. Their practices appealed the teacher's mind. Impressed by the sincerity of the disciples, the Dharma Teacher instructed them in the authentic way.

The instruction was as follows: "Thus calming the mind. Thus practicing the way. Thus coexisting with the world. Thus acting. This is how to calm the mind by the Mahāyāna Way. Please do not make mistakes." *Thus calming the mind* refers to the wall-meditation (or *dhyana* meditation). *Thus practicing the way* refers to the Four Practices (to be given below). *Thus coexisting with the world* refers to protection from criticism and blame. *Thus acting* is to let the action free itself from attachment. Here I have described briefly how it happened and what was the teaching. The details are as follows.

(b) *The main text: Two Entrances and Four Practices*

There are diverse ways to enter the Supreme Path. However, I would say that they are summarized into two, not more. One is (A) to enter by immersing into the universal principle. The other is (B) to enter through practice.

I. Entering by the universal principle

Entering by the universal principle is to immerse oneself into the universe by realizing the fundamental principle according to the Teaching. In principle, ordinary persons and saints (among the sentient beings) are the same in their true nature, but one is unable to see the true nature clearly because one's mind sight is clouded by external dusts, and thus deluded. Having deep insight into this aspect is essential. If one throws away the delusion and immerses oneself into the true nature, if one settles constantly in wall-meditation (or dhyana meditation), if one firmly maintains the state of no discrimination between self and other, and further if one is free from written teachings, then one is unified harmoniously with the universal principle. Without fabricating any view, one is immersed in a universal, natural, and settled state that is devoid of intentional action. This is called *entering by the universal principle.*[3]

II. Entering by practice

Entering by practice refers to the "Four Practices." All of the others are included in these practices: (a) *practicing patience*; (b) *practicing to mindfully accept surrounding conditions*; (iii) *practicing liberation from seeking*; and (iv) *practicing the Dharma.*

(a) *Practicing patience*
When a person practicing the Supreme Way suffers hardship, they should recite: "In the past, I have forsaken what is fundamental and pursued

---

[3] *Remark by the translator* (T. Jixin Kambe) on *entering by the universal principle:*
This is in accord with the Mahayana concept of *Buddha Nature.* Entering by the fundamental principle is to immerse oneself into the fundamental function as well as internal *self-evolving physiological* function existing within oneself and interacting in harmony with the surrounding world. More explicitly, Bodhidharma's path is to realize it by cultivating the self-possessing Buddha Nature. By the power of dhyana meditation or Zen meditation, the internal mind is calmed, then becomes serene, and shines like a bright moon.

frivolously dissipated life for a very, very long time (a number of *kalpa*s). During transmigration through various states of existence (i.e. during the samsara wandering), I incurred others' anger or others' hatred an infinite number of times, and also caused harm to others. Though I am no longer committing such dark deeds, my present suffering is the ripened effect of the accumulation of misdeeds. These are neither caused by humans nor by heavenly beings. With patience, I just swallow all of these and neither complain nor accuse others." In a sutra, it is said: "*When confronted with hardship, don't worry about it and just accept it. Why? Because you have settled yourself at the fundamental level of awakening.*" When one practices with this heart, one is pacified and purified, i.e. immersed in the universal principle. The person goes ahead on the way, bitterness accompanied with patience. This is called *practicing patience.*

(b) *Practice of mindfully accepting surrounding conditions*

Sentient beings are *without self* and hence exist only in relation to surrounding conditions. Since they all arise from surrounding conditions, they all have an equal change of experiencing bitterness and happiness. Even if we have good luck and honor, it is due to our past karma and we only enjoy its effect now. But if the conditions run out, it shall be gone. Why on earth should one feel joyous about them? Thus, gain and loss arise from conditions. One should not waver between hope and despair, nor be moved by joyful transient winds. Then, under the power of the universal principle, the Way is perfected with mindful practice. This practice is called *mindfully accepting surrounding conditions.*

(c) *Practice of non-seeking*

Ordinary people are always deluded and seek something else here and there. This is called *seeking*. Wise men and women awaken to the Truth and aspire from their heart to stay away from the worldly chasing. They pacify themselves by adopting a non-seeking mindset. Body and form change continually according to changing conditions, and nothing stays unchanged. Hence everything is void of fixed entities. There is nothing to seek or desire.

Consider the following episode from a sutra. The goddess of *good luck* visited someone's house and said that she coud give the family precious jewels. They were pleased. Subsequently, a sister goddess of *darkness* visited them and said that she could extinguish all the properties. Both *good luck* and *darkness* always go together and follow after each other. Reflecting ourselves, we always encounter such cases when living in the samsara. It is as if we were living in a burning house. A sutra says, "Seeking anything always results in bitterness. Seeking nothing lead to peace and pleasure." Thus, it is clear that "to seek nothing is to truly practice the Way."

(d) *Practice of the Dharma Way*

If a phenomenon is clean and pure in nature, it is named as Dharma. The Dharma is devoid of any aspect to be characterized: neither taint nor clinging, nor this nor that.

[*Translator's note*: *Practicing the Dharma* is to realize the truth of the pure Path, that *all forms perceived are empty in nature*. In fact, everything in the world is devoid of self-entity.]

The sutra says, "The Dharma has nothing to do with the sentient beings because it is free from the taint of sentient beings. There is no Self in the Dharma because it is detached from the taint of Self." If wise men and women trust the truth and realize what it means, then they should practice the Dharma precisely. The Dharma is free from stinginess. One who practices the Dharma should offer one's own body and possessions and practice *dāna-pāramitā*. Without any stinginess in mind, and with a deep understanding of the three emptinesses (of donor, receiver, and alms), and without sticking to anything, practice only for liberation from tainted mind and to save beings from suffering, but do not stick to what has been done. Thus, both self and others are saved and profited. This adds the grace to the grandeur of the Way of Awakening and it is nothing but practicing the Dharma.

This is *dāna-pāramitā*, the first of the six *pāramitā*s (see Section 8.6, referred to as bases of practice for the Dharma). The other five are like it. For liberation from delusions, one practices the six *pāramitā*s and nevertheless does not leave the trace of practice behind. This is the practice of the Dharma.

Epilogue

More than two thousand years ago, there was an encounter between Western Greek philosophy and Eastern Indian Buddhist philosophy in the ancient Indo-Greek kingdom at Bactria, which covered Afghanistan and northern Pakistan, mainly north of the Hindu Kush mountains stretching into Pakistan. There were Greek communities in the Bactrian area and also in northwestern India. As mentioned in the Prologue and Chapter 1, the Indo-Greek King Menander I of Bactria was interested keenly in Buddhism and met the Buddhist sage Nāgasena. They had sincere intellectual dialogues in the 2nd century BCE. This was essentially the first encounter between East and West philosophies in human history. It was recorded in a famous Buddhist text, the *Milindapañha*, Questions of Milinda (Menander).

More than two thousand years since then, we are now witnessing a renewed encounter between the two streams that have been revolutionarily transformed from their ancient ancestors. Chapter 11 described the scientific revolution that occurred in European countries in the pre-modern age from the 16th to the 18th centuries, modernizing ancient Greek natural philosophy. It was a real turning point in the sciences studying nature, after which the sciences evolved into their modern forms.

Now, in the 21st century, it is amazing to learn that modern sciences have recently provided us with scientific evidence supporting Buddhist philosophy. The brain sciences are concerned with no-self nature of human existence [40]. There is nothing in the structure of our brain that can be identified as a self. This was considered in detail in Chapter 8. Another, more fundamental point is described further below. Very recent studies by a group of scientists imply that the primordial Earth had sufficient molecular materials and gases necessary for the creation of a biosphere (life). This approach has disclosed new biological principles in addition to the principles of physics and chemistry. The biosphere is explained by two principles: biodiversity and coexistence among diversely different species. This may imply a primordial aspect of Buddha-nature.

Buddhists' Approach to Nature is Scientific

The main theme of the present work is whether there is a common platform between science and Buddhism. To approach the reality of world, both scientists and Buddhists take the same stance: scientists study the reality and diversity of nature and the dynamical evolution of the world, while Buddhists are concerned with the realistic and natural interrelation between themselves and the world surrounding them, with the goal of realizing a *calm* and *peaceful* mind within themselves. One may say that both are taking *scientific* approaches from two different sides, just like approaching the top of a high mountain from two different routes: that of the intellect[1] and that of wisdom. From the Buddhist perspective, *wisdom* is timewise intelligence and compassion (Section 5.6). There is a passage in the Buddhist text *Nirvana Sutra*[2]:

> "*If you wish to know what the Buddha-nature is, listen carefully,*"
> "*one must think of time and materials*[3] (and their transformations).*"
> "*All beings possess the Buddha-nature as in the nature of butter which is in milk.*"

It is seen that Buddhists' approach to nature is common with that of scientists. In particular, look at the statement "*Think of time and materials,*" which is consistent with physics and chemistry. In Chapter 2, we saw the ancient scientific philosophy from the *Abhidharma-kośa-bhāṣya.* In Chapter 3, too, we learned that most observations of the ancient *Abhidharma* philosophy have features in common with those observed by modern brain science.

---

[1] Intelligent data of information are limited in time by a time horizon (see below), beyond which initially close data are no longer close.

[2] The reference [64] (b), Chap. 34, On Bodhisattva Lion's Roar (b).

[3] In Buddhist terms, materials are expressed by the terms *form* and *colour*, in the BCE ages.

Apart from what they learn from school textbooks, people do not know the long, long history of biological evolution because they have neither personal nor direct experience of it. Moreover, they know neither about the geological ages nor about cosmological evolution from self-experinecce. They get that information only from textbooks, TV education programs, or other media. They are informed only in terms of scientific knowledge. Not only that, but because human beings have no memory at all of being in their mothers' wombs, there is no way to know their ancestors beyond time. Hence it is unreasonable to say that every human being is well versed in everything. One cannot claim that one has clear knowledge of everything and behaves appropriately. It is said in Buddhism that every human being is intrinsically in the grip of *ignorance*, *nescience*, *avidyā*, or 無明 (Chap. 4), as expressed in various languages. *Ignorance* is explained as a *bondage*, a *proclivity*, interpreted as a sense impelling the soul towards external objects (Section 4.2), provoking careless attachment. This causes humans to fall into unwanted troubles. Attachment is a driving force in the formation of karmic action (Section 2.5).

Such physiological mechanisms that impel ourselves (mind and body) to external objects exist inherently within our living body and brain. It is a latent defilement that is difficult to remove. Under ignorance or nescience, one is constantly followed by a sequence of such physiological mechanisms. This is *beginningless* (Section 5.2). Consider the stages of plant growth. Plants go through several stages, including vegetative, budding, flowering, and seed production. Plants produce seeds from another seed. As a result, the cycle of a living being has no beginning. Similarly, in terms of ourselves, we are trapped in an endless cycle of existence. All phenomena (dharmas) arise as a result of causes under certain conditions. The causes are themselves dependent on other causes. This implies that the chain of causes has no beginning.

Scientific Aspects of the Biosphere on Earth

Now, in the 21st century, using modern science, let us try to contemplate the notion of the beginningless cycle of our existence. The age of Earth is considered to be around 4.5 billion years. According to fossil evidence of a microorganism that was discovered recently by Japanese scientists in old

Canadian rocks, the earliest life form on Earth might be about 3.9 billion years old. At that time, the sea had already formed. The time required for first life, in the form of single-cell organisms, to emerge from the beginning of the Earth was 0.6 billion (600 million) years. The time interval of 600 million years is sufficiently long because it is longer than the time interval between the present and the first emergence of land animals evolved from sea animals (such as lobe-finned fishes), which occurred only about 370 million years ago. Land animals such as *amphibians, reptiles, birds,* and *mammals* evolved over 370 million years.

Imagine the Earth to have been like a flask for chemical reactions. The story [96] by a group scientist concludes that, during such a long time of 600 million years, the Earth Flask succeeded in creating the first life of single-cell organisms from the primordial (early) Earth atmosphere and sea water. According to geological scientists, the primordial atmosphere was formed by outgassing from the solid Earth, implying that the Earth's original atmosphere was mostly composed of water (H_2O), carbon dioxide (CO_2), and nitrogen gas (N_2), but it lacked free oxygen. For oxygen to become a significant component of the Earth's atmosphere, another 1.5 billion years were required, until tiny organisms known as *cyanobacteria,* or *blue-green algae,* conducted photosynthesis using sunshine, water, and carbon dioxide.

Intense volcanic activity released gases that made the early atmosphere. Volcanic activity also released water vapor, which condensed to its liquid state, water, as the Earth cooled, forming the oceans. Nitrogen was probably also released by volcanoes and gradually accumulated in the atmosphere. Magma was rich in iron (Fe). Very occasionally, Earth gets bombarded by large meteorites, as well as space dust and *micrometeorites* made of iron.

The origin of life on Earth is currently being investigated worldwide by scientists in various fields on the basis of scientific experiments and observations. Some of them (mainly scientists in Japan) are proposing an amazing scenario of how life began on the primordial Earth. Their findings from careful and patient investigation in various fields are as follows. The Earth system acted like a huge flask of chemical reactions synthesizing unicellular organisms that thrive at relatively high temperatures.

The emergence of the first life form on Earth necessitates the presence of two essential macromolecules, proteins and RNA, which has a blueprint to produce protein molecules. In general, amino acids are the building blocks of protein. More than 500 naturally occurring amino acids are known to constitute proteins. However, all life on Earth is based on 20 amino acids, which are governed by DNA to form proteins. It is known that the meteorites bombarding from the sky include 11 of these amino acids. Scientists have disclosed that remaining amino acids can be produced in a laboratory experiment from water, CO_2, N_2, and Fe existing in the primordial atmosphere using a high-speed projectile that simulates meteorite bombardment on the sea surface. This is really an amazing step forward in the theory of the origin of life on Earth.

Consider the building blocks of RNA, which is another essential element for the initial life. RNA is a polymer made of long chains of *nucleotides*. A nucleotide consists of a sugar molecule (ribose) and four nucleotide bases. The four bases of RNA are *adenine* (A), *cytosine* (C), *guanine* (G), and *uracil* (U). Recent laboratory experiments have made it clear that those four bases of RNA can be formed in the environment corresponding to the primordial Earth.

Summing up, (i) the primordial Earth could be a *flask* for chemical reactions. Not only were the four bases of RNA synthesized naturally there, but also the RNAs; (ii) the macromolecules consisting life form were synthesized chemically; (iii) the Earth Flask did it.

Thus, the primordial Earth was a huge scientific laboratory with hot sea water and ground rocks that was exposed to abundantly bombarding meteorites. So far, it has been shown that the Earth Flask is equipped with the necessary materials to synthesize the macromolecule RNA and sufficient amino acids to make protein molecules.

Now we must remind ourselves that RNA has a blueprint for producing protein molecules. In order to produce new protein molecules, another molecular device is used to read the blueprint sequences written on RNA. *Ribosome*s, which are macromolecular machines that perform biological protein synthesis (RNA translation), are the device. Having carried out computer tests for a set of 10^{12} proteins, a Japanese scientist discovered *ribosome* proteins, which become a translational apparatus by integrating with RNA.

Thus, from the studies of scientists, it is seen that the primordial Earth had sufficient materials that were able to reproduce molecules necessary for life. This has a profound philosophical meaning for us living on Earth today. Moreover, it is really remarkable to know that this approach has disclosed a new biological principle in addition to the principles of physics and chemistry. The *biosphere* is explained by the two principles: *biodiversity* and *coexistence* among diversely different species. This reminds us of Buddha-nature.

It is known in physics that time evolution of a physical system composed of diverse species shows complex non-trivial behaviors, called *chaotic* behaviors. One of the properties to be noted here is that predictability of physical evolution of the system is limited in time by a time-horizon,[4] beyond which the system behavior cannot be predicted precisely, rendering long-term physical prediction of its behavior impossible in general. Future states are even more unpredictable when the system is exposed to external disturbances or uncontrollable forces. This opens the long-awaited opportunity for the emergence of a new *biological principle* in addition to the principles of physics and chemistry. The next section gives an account of *coexistence.*

Thus, it might be concluded that a primordial biosphere brewed up in the Earth Flask from the primordial Earth system and that *biodiversity* and *coexistence* were created in the primordial Earth. It would not be a mistake to say that the property of biodiversity and coexistence is a primitive sprout of Buddha-nature.

The Beginninglessness of Our Existence

Important and profound information on the life of Earth has been given in the last paragraph. Early life was formed chemically and physically in the primordial Earth, which was covered with primordial atmosphere and sea.

---

[4] The time-horizon depends on physical systems under consideration. In our daily lives, weather prediction would be reliable for two days, at most one week. The Earth's climate change is project to last for hundreds of years. Chaotic oscillations of chemical or hydrodynamic systems in the laboratory are of the order of seconds or minutes. However, the Solar System is estimated to remain in its present form for another five million years. Concerning the origin of life, 600 million years are proposed for its synthesis within the Earth's flask.

Those surface layers were composed of the main components CO_2, N_2, and minerals, as well as many other molecules that had no direct bearing on the formation of life. But their existence might be essential for the primordial surface layers. The primordial Earth was exposed to abundantly bombarding meteorites.

Once macromolecules such as RNA and ribosomes were formed, they could produce or reproduce necessary molecules by themselves. After a long time (beyond the predictability horizon of physical laws), they eventually formed a biosphere under the biological principle of *biodiversity* and *coexistence* among countlessly different species. Biodiversity means that the system is composed of countless species and is constantly evolving. In other words, the system does not stay in a fixed state, but its countless species interact with one another to maintain biodiversity. The constituting members of the system were governed by the principle of coexistence, according to which each member kept changing but never went out of existence. The computing model mentioned above has shown that when one species became extinct, others were influenced significantly and the system was in general eventually destroyed as a whole. This is conceivable as nonlinear behaviors of systems consisting of countless species interacting with one another.

In the system evolving in the way described above, each member is changing and has no intrinsic nature. This kind of existence may be characterized by emptiness. The biosphere evolved from the primordial Earth atmosphere, which was formed from countless astronomical events preceding it. Hence, they are said to be *beginningless*, and our existence is also beginningless and characterized by *emptiness*. One may ask when the consciousness of biological beings emerged. The Earth system itself might have had primordial consciousness and primordial wisdom. The early biosphere might have already had primordial Buddha-nature.

Imagine a *Buddhist eagle* soaring in the sky of *emptiness* with both wings spread out, one wing of *wisdom* and the other of Great Compassion for the coexistence of all countless sentient beings.

Brief Note on the Big Bang Theory

Concerning the beginning of our universe, one must mention the Big Bang theory, which is well known in modern science. The Big Bang theory [97] is the prevailing cosmological model explaining the observed universe represented by the Hubble–Lemaître law. This law states that the farther away a galaxy is, the faster it is moving away from Earth. Extrapolating this cosmic expansion backwards in time, the theory describes an increasingly concentrated cosmos, preceded by a singular state in which space and time lose meaning. Hence modern science is unable to predict the state. Detailed observations of the universe place the Big Bang singularity at around 13.8 billion years ago, considered the age of the universe. No valid physical law is known about the singularity of very high density and temperature. Needless to say, the time scale of 13.8 billion years is much longer than 0.6 billion (600 million) years, which was mentioned above as being sufficiently long.

Extrapolation of the expansion of the universe backwards in time using the general relativity theory yields an infinite density and temperature at a finite time in the past. This singular behavior, known as the gravitational singularity, indicates that general relativity is unable to give an adequate description of the laws of physics in this regime. Even in modern science, its beginning is not described by any valid theory. Furthermore, the Big Bang theory describes only the universe to which our Earth belongs. The theory cannot predict the existence or non-existence of other universes.

The following is an informative remark. The Big Bang is not an explosion in space, but rather an expansion of space, which only the general relativity theory can describe. In other words, matter is not moving outward to fill an empty universe. Instead, space itself expands with time everywhere and increases the physical distances between comoving points. As a simple example, consider making a raison bread by baking. Baking a dough mixed with yeast and a number of raisons causes it to expand after a while. Then the intervals between neighboring raison particles expand. In this model of the expanding universe, the expanding raison particles represent the expanding galaxies.

Modern age of Buddhist Schools

Shakyamuni Buddha expounded what is called the Middle Way, avoiding two extremes. After several hundred years since the time of Shakyamuni, a deep philosophical concept was proposed by Ācārya Nāgārjuna, who founded the philosophy of *śūnyatā* (*emptiness*) or equivalently the Middle Way. This activated traditional Buddhism and encouraged new movements, stimulating the emergence of various new Buddhist schools: *Gandharan* Yogācāra (4th century: Asaṅga and Vasubandhu brothers), Chinese Chan Buddhism (6th century:Buddhists, Bodhidharma, and HuiKe), and Indo-Tibetan Buddhism, including *Vajrayāna* (8th century: Indian Buddhists Padmasambhāva and Śāntarakṣita and the Tibetan Yeshe Tsogyal).

In the modern era, Buddhism is experiencing another renewal of internationalization from its past locales of South and East Asia. The newly emerged Buddhism is the Buddhism spoken about in European languages and US-American languages,[5] although traditional Buddhist schools are still active in every country of East and South Asia.

By the initiative and the leadership[6] of Ven. Thanat Inthisan (Ph.D.) and Ven. Katugastota Uparatana in 2015, the International Buddhist Association of America (IBAA) was established in February 2017 to promote Buddhism, conduct spiritual worship, Buddhist studies, and perform religion activities in a spirit of non-sectarian tolerance of Buddhist communities in the United States. The IBAA has been officially accepted as a non-profit religious organization in the United States.

In June 2016, on the occasion of the 40th anniversary of the Council of Thai Bhikkhus in the U.S.A., an international Buddhist conference was held at Wat Nawamintrachutis, Cambridge–Boston, organized by Ven. (Dr.) Thanat Intisan (Secretary General of the Council of Thai Bhikkhus in the U.S.A.), to which the author (T. J. Kambe) was invited as one of the speakers.

---

[5] Correspondingly, at the time of first encounter in the BCE era, it might be characterized as the Buddhism represented by Indian languages, or Sanskrit, or Pali, and later by Chinese languages.

[6] *Bhante Katugastota Uparatana Maha Thera* is the Chairman of the IBAA. He was ordained as a Buddhist monk in 1966 in Ranwalagedara Viharaya, Sri Lanka. Since 1982, he has served the Washington DC community as a monk, teacher, and Vipassana meditation teacher. Bhante Uparatana is a founder of The International Buddhist Center; Venerable Phramaha Thanat Inthisan, Ph.D., is the Vice Chairman of the IBAA and Secretary General of the Council of Thai Bhikkhus in the U.S.A.

One more aspect must be emphasized. Modern Buddhism is being revived by Tibetan Buddhists, who have been spreading it outside Asia since 1959. In particular, without the precious work of the edition [2][7] by Venerable Gelong Lodro Sangpo in 2012, it is certain that Chapter 2 of the present book would not have come into being. The same can be said of Chapters 4, 5, and 6.

Walking in the Valley of the Buddha

The *Buddha Carika*[8] or *Dhammayatra* (Dharma Walk) has been held in every December since 2014 along the Jethian Valley path, close to the ancient town Rajgir, Bihar, India. The path route (Ep-1) connects King Bimbisara's town Rajgir with the Bodh Gaya, where Gautama Buddha had attained *bodhi*. This path follows one of the ancient paths taken by the Buddha himself during his life in various parts of India. Today, Rajgir[9] is surrounded by low-lying picturesque hills (Ep-2). To protect the forest there, an area of 35 was registered as the Rajgir Wildlife Sanctuary[10] in 1978 under the Nalanda district administration.

This walk was first orgnaized in 2014 by several Buddhist organizations intending to revive and spread the history of the Jethian–Rajgir valley. Before beginning the walk, Jethian village people gather for a meal offering to monks and Buddhist participants in observance of *Sanghadana*, an ancient tradition that stems from the time of Buddha, who, while staying in the village, would visit homes to collect food.[11]

---

[7] *Abhidharmakosa-Bhasya* of Vasubandhu (The Treasury of the Abhidharma and Its (Auto) Commentary (ca. 380–390 CE), translated by L. de La V. Poussin (originally 1823, 1971).

[8] Cārikā in Pali means a journey, or wandering.

[9] Rajgir (historically known as Rājagṛiha) means "the City of Kings," an ancient city and a municipal council in the Nalanda district of the Magadh region in the Indian state of Bihar. The city of Rajgir (Sanskrit: Rājagṛha; Pali: Rājagaha) was the first capital of the kingdom of Magadha, a state that would eventually evolve into the Mauryan Empire.

[10] See the next section, "Fujii Guruji," for the environmental policy on the World Peace Stupa at Rajgir.

[11] The Jethian Valley — Walking in Buddha's Foot Step. https://passportandbaggage.com/the-jethian-valley-walking-in-buddhas-footsteps/

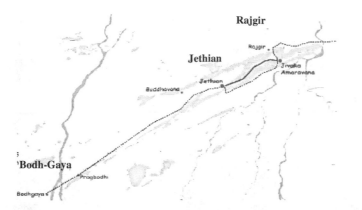

Ep-1. The walking path from the Jethian village to Jivaka Amaravana. https://lbdfi.org/portfolio-items/jethian-valley-2014/

Ep-2. Walking in the Jethian Valley (Rajgir Wildlife Sanctuary).

In the afternoon, the opening ceremony (Ep-3) was held by Buddhist monastics and devotees from around the world, faculty members from Nava Nalanda Mahavihara and Nalanda International University, the Bodh Gaya Temple Committee, scholars, heritage conservation specialists, and representatives from the village [98]. Several speakers highlighted the rich cultural significance of the ancient Magadha region, and the importance of preserving the historic and

Ep-3. IBAA President Wangmo Dixey delivering the opening Address. The Most Venerable Uparatana (Chairman of IBAA) is sitting fourth from the right.

Ep-4. Walking, living Sangha.

religious integrity of the Jethian Valley in an authentic way for current and future generations.

Wangmo Dixey (President of IBAA)[12] said the following in her address:

> *"We have come here to follow the footsteps of the Blessed One (Buddha) and it was a highly joyous and sacred moment for us to see the **Sangha-dana** being performed in the presence of the **living Sangha**. That we can work harmoniously together, even though we come from ten countries and share many different languages, is a testament to the unity of the Maha-sangha, a shining example of our commitment to the teachings of the Buddha Bhagavat."*

The area of the Jethian Valley has been kept preserved under the national policy of India, as noted in the next section ("Fujii Guruji").

On December 13, 2017, the present author was honored to join in the *Buddha Carika*. The Dhammayatra began from the village of Jethian at around 3 p.m. Amid chants of *"Buddham sharanam gacchami; Dhammam saranam gacchami; Sangham saranam gacchami,"* over 500 pilgrim walkers started the 15 km walk through the forest trail in the pristine environment and "untouched" beauty of the Jethian Valley (Eps-2, 5, 7, 9).

During the ancient Buddha's time, Rajgir was the capital of the Magadha Kingdom. According to legend, King Bimbisara of Magadha had come and greeted the Buddha in this Jethian Valley and offered Gautama Buddha the parkland of *Velu-vana* (bamboo grove), surrounded by a bamboo (*velu*) forest (*vana*). The King chose this place for the

---

[12] Wangmo Dixey serves as the elected President of the IBAA (*International Buddhist Association of America*) which brings together all three Yanas (Śrāvaka-yāna, Pratyekabuddha-yāna, Bodhisattva-yāna) into one platform as a voice for them in the United States. She serves as Executive Director of LBDFI (Light of Buddhadharma Foundation International) based in Berkeley CA. For ten years, LBDFI has worked to restore the Buddha Sasana in India.

Ep-5. Walking in the Jethian Valley.

Ep-6. Veluvana Vihara.

Ep-7. Vulture Peak.

Buddha since it was a quiet spot suitable for meditation, as well as a convenient location not too far outside the town. Since then, Buddha used this place as a Buddhist vihara and spent the second, third, and fourth rain-retreats at Veluvana Vihara.[13] The destination of the Walking was the Veluvana Vihara (Ep-6).

In nearby mountain area, there is the Gridhrakuta peak (Vulture Peak, Ep-7). In *The Lotus Sutra*, this place is described as follows: "Thus have I heard. Once the Buddha was staying in the city Rājagṛha, on the mountain called Gṛdhrakūṭa, together with a great assembly of twelve thousand monks" (Chap. 1 [100]). In Chap. 5 of the same sutra: "The Tathāgata has immeasurable, unlimited, and incalculable merits." Moreover, "Tathāgata

---

[13]Vihāra is the Sanskrit and Pali term for a Buddhist monastery. It originally meant "a secluded place in which to walk" and referred to "dwellings" or "refuges" used by wandering monks during the rainy season. The northern Indian state of Bihar derives its name from the word *vihara*, due to the abundance of Buddhist monasteries in that area.

Ep-8.   Monkes of the Nippon-Myohoji.

knows the character of liberation, dispassion, cessation, complete nirvāṇa, and eternal tranquility which ultimately leads to emptiness."

Buddha's mind was always calm. He had attained ultimate, eternal tranquility. The photo in Ep-9 shows the view from the peak, which is calm and marvelous. In fact, the surrounding atmosphere was just exhibiting the eternal tranquility of Buddha's mind in front of us.

At the Vulture Peak (Ep-7), the Chanting Ritual was held after *Walking for Peace* in 2018 by the Buddhist participants from various countries, such as Tibet, Nepal, and Japan. Japanese monks of the Nipponzan-Myohoji school are sitting in the front row of the photo Ep-8. They are disciples of the spiritual leader Fujii Guruji (see the next section). They chanted *"Na Mu Myo Ho Ren Ge Kyo"* by beating a hand-held prayer drum (placed in front of a monk, Ep-8) in the calm surrounding atmosphere and uniting the minds and hearts of all people. The chant *"Na Mu Myo Ho Ren Ge Kyo"* is translated literally as "Homage to the *Lotus Sutra*" or "Praise to the Wonderful Dharma of the *Lotus Sutra*,"[14] but followers of the Myohoji think that the chant is ineffable and unable to be

---

[14] The chant *"Na Mu Myo Ho Ren Ge Kyo"* is also understood as the name of the eternal Buddha.

Ep-9. View from the Vulture Peak.

expressed literally; its meaning reveals itself effectively to practitioners through time. Their teacher Fujii Guruji once visited the Gandhi Ashram in Wardha in 1933 (nearly 90 years ago).

Fujii Guruji delighted at seeing Gandhi embracing the drum and attempting to learn the chant *"Na Mu Myo Ho Ren Ge Kyo,"* which was used at the Gandhi Ashram in Wardha during daily prayer. Gandhi later wrote to Fujii Guruji: "The subtle and profound sound of your drum resonates in my ear."

The serene atmosphere at the Vulture Peak is nothing but the Eternal Buddha. In the midst of the surrounding mountains, the chanting by Tibetan monks echoed sublimely at the Peak (Ep-10). Subsequently, we had a moment of bliss in hearing the musical chanting of Nepal Buddhists (Ep-11), soaring in the peaceful breeze.

Just at this moment, the sun was sinking in the westward mountain ridge. It is said that Buddha himself used to immerse in meditation at dusk (possibly at Rajgir dusk).

Not far from the Vulture Peak, there is a site of Buddhist caves about 2 km southwest from Rajgir. That is the Saptparni cave, where, it is said,

Ep-10.   Tibetan monks gracefully and joyfully passing prayer beads (*japamala*, rosary).

Ep-11.   Nepal Buddhists relaxing after harmonious, beautiful chanting.

Ep-12. At Nalanda University, ancient monastic university.

Buddha spent some time before his *parinirvana* and where the first Buddhist council was held after it. After the First Council in the year of Buddha's nirvāṇa, the monks who had taken part in the Council retired to rest at Veluvana Vihara (Ep-6).

On the last day of our stay at Rajgir in 2017, we paid a visit to Nalanda University, the renowned Buddhist monastic university in ancient Magadha, where we had a session of the modern version. In the photo Ep-12, we see (from the right) Ms. Wangmo Dixey (President, IBAA), Ms. Prasarn Manakul, Ven. Dr. Thanat Inthisan (Vice Chairman, IBAA), T. Jixin Kambe (former Professor, University of Tokyo), a Buddhist monk (Council of Thai Bhikkhus in the U.S.A.), Dharmacharya Shantum Seth (Ahimsa Trust, India), and Mr. Prasarn Manakul (General Secretary, IBAA, Wat Thai DC Board).

Dharmacharya Shantum Seth is an ordained Dharma Teacher (Dharmacharya) of Indian origin in the Buddhist Mindfulness lineage of Zen Master Thich Nhat Hanh and teaches in India and across the world. He is a co-founder of Ahimsa Trust and has been a student of Master Thich Nhat Hanh for the last 35 years. He has been leading pilgrimages, including "In the Footsteps of the Buddha" and other transformative journeys, across diverse regions of India and Asia since 1988.

The present author (T. J. Kambe) would like to express thanks from the bottom of his heart to Dharmacharya Shantum for guiding us on our pilgrimage to the ancient Footsteps of Shakyamuni Buddha in 2004. After the tour, it was gracious of him to present the author with a precious book of Master Thich Nhat Hahn, *Old Path White Clouds* [6].

Ep-13. Dharmacharya Shantum Seth.

The 2017 international Buddhist gathering in Bihar, India, drew nearly five thousand participants from eleven countries, all of whom volunteered to travel from their home countries to pay respect to the Buddha Country and Buddha Sassana. This presents a model example of the *modern aspects of Ratna-Traya*: *Buddha, Dharma, Sangha.*

Fujii Guruji:[15] *Rajgir, quest for peace, and never despising others*

"It is the true law of nature that gives birth to life and nurtures heaven and earth."

This is the message emphasized by Fujii Guruji at the end of his article [100], composed at the occasion of Jawaharlal Nehru Award in 1978 from India.

The idea of construction of the **World Peace Stupa at Rajgir** was first conceived in 1956 by Ven. Nichidatsu Fujii Guruji (the Nipponzan-Myohoji school 日本山妙法寺) during his fifth visit to India. According to his article "Seal of Fearlessness" [101], he had sent his proposal in 1958 for construction of the Stupa at the Ratnagiri hill (Rajgir) to Prime Minister Jawaharlal Nehru and President Rajendra Prasad. They agreed to the proposal, and the construction of the Stupa was completed in 1969. The foundation stone of the first Shanti Stupa was laid by Pandit

---

[15]The Most Venerable Nichidatsu Fujii (1885–1985), commonly known as Guruji [100].

Ep-14. Fujii Guruji (age 67) standing with a drum in front of the first Peace Pagoda in Japan, 1952. [100]

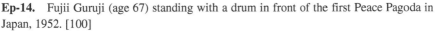

Jawaharlal Nehru at Rajgir, and it was built in marble on the Ratnagiri hill. Thus, the Vishwa Shanti (World Peace) Stupa was constructed by the Nipponzan-Myohoji sect in 1969 on the Ratnagiri hill. The hill envelops the Griddhakuta (Vulture Peak). Rajgir is surrounded by five hills, which have been preserved by the Indian government's policy to be untouched, scenic, and serene. The Buddha's abode (for meditation) is known as the sixth hill — Griddhakuta (Vulture Peak). In 1952, the first Peace Pagoda was constructed by Fujii Guruji in Japan as a symbol of peace (Ep-14).

In 1933, Monk Fujii met Mahatma Gandhi while staying at his ashram in Wardha, Maharashtra. According to the article [101], "Unfortunately I could neither speak English nor Hindi. I therefore submitted an English translation of my views to Gandhiji. This is how it came to pass that I found residence in the Wardha ashram." Thus, he was granted permission to stay there. Gandhiji (as Fujiiji called him) was honored by his presence and added the *daimoku* chanting *"Na Mu Myo Ho Ren Ge Kyo"* to the ashram's prayers. Mahatma Gandhi was the first to address Monk Fujii as "Guruji" during the latter's stay at the Wardha ashram in 1933. Since then, throughout his life, he was so addressed affectionately and respectfully by his many foreign followers and admirers. Guruji later recollected: "The Pacific war raged ever more brutally. I could no longer keep silent about the war, in which people were killing one another. Thus, I traveled through the whole of Japan and preached resistance against the war and advocated the prayer for peace. It was a time in which any person who only spoke about resistance to the war, would go to prison because of that alone."

Having seen the destruction of Hiroshima and Nagasaki by atomic bombs, Guruji undertook the task of building world peace pagodas through absolute nonviolence. For the next forty years of his life, his peace pilgrimage took him to different parts of the world — in both the East and the West.

Guruji wrote in [101] as follows:

Immediately following the Independence in 1947, Indian leaders gathered at the Sanchi Stupa and determined the principle of the policy of India, which was to establish a peaceful nation. A resolution was passed to restore Buddhism, a religion of peace, in the interest of building a peaceful nation. As an initial step to restore Buddhism, a committee to reconstruct Rajgir was organized in 1956. Prime Minister Nehru himself headed the committee and five members were appointed. Despite my being a foreign national, I became a member of this committee in my capacity as a disciple of the Buddha. The erection of the Rajgir stupa and a stupa in Kalinga are some of the works by this committee to restore Buddhism.

Thus, the Rajgir Wildlife Sanctuary (Ep-2) is the updated version of the Indian principle in the interest of protection of forests and was notified as Pant Wildlife Sanctuary, Rajgir, in 1978 by the Nalanda Forest Division, District Administration, Bihar.

Guruji loved to tell the story of Sadāparibhūta Bodhisattva (the Bodhisattva who never despises). According to the *Lotus Sutra* [99], the Bodhisattva would always pay obeisance to others, saying, "I profoundly revere you all. I dare not despise you. Why? You are all treading the path of bodhisattvas and will in time succeed in becoming buddhas." In the *Lotus Sutra* [100], Chapter XX, there is the following passage:

> At that time there was a monk, a bodhisattva, called Sadāparibhūta (Never Despising). Why was he called Sadāparibhūta? Because when-ever he saw any monk, nun, layman, or laywoman, he would praise and pay homage to them, saying:
> "I deeply respect you. I dare not belittle you. Why is this? Because all of you practice the bodhisattva path, and will become buddhas."

With Gassho,
Na Mu Myo Ho Ren Ge Kyo

References

[1] *Milindapañha: The Questions of King Milinda*, translated by T. W. Rhys Davids (1890), reprinted in 2012 (The Sacred Books of The East vol. 35 and 36; edited by F. Max Muller).

[2] *Abhidharmakosa-Bhasya of Vasubandhu [AKB]: The Treasury of the Abhidharma and Its (Auto) Commentary* (ca. 380–390 CE), translated by L. de La V. Poussin, G. L. Sangpo, Motilal Banarsidass Publ., 2012 (originally 1823, 1971); *Vasubandhu*, Internet Encyclopedia of Philosophy, K. T. S. Sarao (Delhi University), 2015.

[3] *Jewels from the Treasury (Abhidharmakosa): Vasubandhu's Verses on the Treasury of Abhidharma*, its commentary, *Youthful Play* by the Ninth Karmapa Wangchuk Dorje, translated by David Karma Choephel, KTD Publications (Woodstock, NY), 2012.

[4] Dharmaraksita, Wikipedia.

[5] Abhidharmakosa Study Blog, https://abhidharmakosa.wordpress.com/ 2012

[6] Thich Nhat Hahn (1991), *Old Path White Clouds*. Indian edition 1997 by Full Circle.

[7] "The Daisetsuzan Mountain Range: Kamuy Mintar", at the site, *Kamikawa Ainu in Coexistence with Kamuy,* https://daisetsu-kamikawa-ainu.jp/en/ story/daisetsuzan/

[8] コマクサ — Wikipedia, https://ja.wikipedia.org/wiki/%E3%82%B3%E3 %83%9E%E3%82%AF%E3%82%B5

[9] "Great Snow Mountain (Daisetsuzan）: Flower Travelogue–Short Summer of *Kamuy Mintar*": TV documentary, 1985, by NHK (Japan Broadcasting Corporation), https://www2.nhk.or.jp/archives/tv60bin/detail/index.cgi? das_id=D0009040222_00000

[10] *Milindapanha* "ミリンダ王の問い" (1) ~ (3), Japanese translation by H. Nakamura, K. Hayashima (Heibon-sha, Tokyo), 1963.

[11] *Milindapanha* "The Debate of King Milinda" translated by Bhikkhu Pisala, Buddha Dharma Education Association Inc. website: www.buddhanet.net; New revised edition, 2001.

[12]　Awakening of Faith in the Mahāyāna (大乗起信論） — Wikipedia.

[13]　Maha-parinibbana Sutta (From Last Days of the Buddha), translated from the Pali by Sister Vajira & Francis Story, (revised edition), (Kandy: Buddhist Publication Society), Copright © 1998 Buddhist Publication Society. 1998. *Japanese edition*: (ブッダ最後の旅」) (Buddha Last Journey, Maha-parinibbana Sutta) by Hajime Nakamura (中村元訳), Iwanami Publ., Tokyo, 1908.

[14]　S. N. Young (2011),　My 21 years with the Journal of Psychiatry and Neuroscience, with observations on editors, editorial boards, authors and reviewers, *J Psychiatry Neurosci*, https://www.jpn.ca/content/36/4/E30; G. Dewar (2016), Empathy and the brain, http://www.parentingscience.com/empathy-and-the-brain.html

[15]　B. Bodhi (1998), *The Noble Eightfold Path*, Buddhist Publication Society.

[16]　C. I. Beckwith (2016), *Greek Buddha: Pyrrho's Encounter with Early Buddhism in Central Asia*, Princeton University Press.

[17]　Yoga-adviser (2016), Abdominal Breathing for Heart Disease, http://yogaadvise.com/abdominal-breathing-for-heart-disease/

[18]　My Interesting Facts World Interesting Facts (2016), 10 Interesting the Lymphatic System Facts, http://www.myinterestingfacts.com/the-lymphatic-system-facts/

[19]　M. Carabottia, A. Sciroccoa, M. A. Masellib, C. Severia (2015), The gut-brain axis: interactions between enteric microbiota, central and enteric nervous systems (review), *Ann Gastroenterol*, **28**, 203–209.

[20]　Healthline (2017), Serotonin: What You Need to Know, http://www.health-line.com/health/mental-health/serotonin#Overview1

[21]　S. N. Young (2007), How to increase serotonin in the human brain without drugs, *J Psychiatry Neurosci*, **32**, 394–399.

[22]　J. Kornfield (2017), Walking Meditation, https://jackkornfield.com/walking-meditation-2/

[23]　G. Dienstmann (2017), Ultimate Guide to Walking Meditation, http://liveanddare.com/walking-meditation/

[24]　A. Sumanasara, A. Hideho (2010), Buddhism and Brain Science (佛教と脳科学) [in Japanese], Samgha (サンガ), http://www.samgha.co.jp; http://www.j-theravada.net/; http://www.serotonin-dojo.jp

[25]　Fumbling Towards Enlightenment, *Peace in Every Step* (Nov. 26, 2010), https://fumblingtowardsenlightenment.wordpress.com/

[26]　P. K. Abraham (2016), *The Mirror Neuron System in Rehabilitation*, A power-point file by a Therapist at Hamad Medical Corp., Qatar.

[27] V. S. Ramachandran, L. M. Oberman (2006), *Broken Mirror: A Theory of Autism*, Scientific American, 63–69.

[28] Mirror Neuron, Wikipedia, https://en.wikipedia.org/wiki/Mirror_ neuron. Sites at Penn State Log In: *Psych 256 Cognitive Psychology SP15*, https://sites.psu.edu/psych256sp15/2015/01/30/mirror-neurons-relation-to-autism/

[29] P. Bhargava (2012), Mirror Neuron, CMSC 828D Report 3, https://www.cs.umd.edu/class/fall2012/

[30] The American Institute of Stress, http://www.stress.org/take-a-deep-breath/

[31] W. J. D. Doran (2016), *The Eight Limbs, The Core of Yoga*, Yoga Art Gallery & Yoga Resources, at the website "Expressions of Spirit", http://www.expressionsofspirit.com/yoga/eight-limbs.htm

[32] M. Carrico (2007), *Get to Know the Eight Limbs of Yoga*, Yoga Journal Yoga for Beginners, http://www.yogajournal.com/article/beginners/the-eight-limbs/

[33] Manas-vijnana, Wikipedia, https://en.wikipedia.org/wiki/Manas-vijnana

[34] Eight Consciousnesses, Wikipedia, https://en.wikipedia.org/wiki/Eight_ Consciousnesses

[35] G. Northoff (2016), *Neuro-Philosophy and the Healthy Mind*, W.W. Norton, New York.

[36] M. D. Fox, *et al.* (2006), Spontaneous neuronal activity distinguishes human dorsal and ventral attention systems, *Proc Natl Acad Sci USA*, **103**(26), 10046–10051.

[37] J. H. Austin (2014), *Zen-Brain Horizon*, MIT Press, Massachusetts.

[38] Y. Ikegaya *et al.* (2004), Synfire chains and cortical songs: temporal modules of cortical activity, *Science*, **304**, 559–564.

[39] G. M. Edelman (2003), Naturalizing consciousness: a theoretical framework, *PNAS*, **100**, 5520–5524.

[40] V. S. Ramachandran (2007), *The Neurology of Self-Awareness: The EDGE 10th Anniversary Essay*, https://www.edge.org/3rd_culture/ramachandran07/ramachandran07_index.html

[41] http://scicurious.scientopia.org/2011/05/04/science-101-the-neuron/;http://science.howstuffworks.com/life/inside-the-mind/human-brain/brain1.htm

[42] L. Jamspal, N. S. Chophel, P. D. Santina (translators) (1978), *Nagarjuna's Letter to King Gautamiputra (Suhrllekha)*, Motilal Banarsidass Publishers, Delhi, India.

[43] J. Walser (2005), *Nagarjuna in Context*, §2, Columbia University Press, Nagarjuna.

[44] Kumārajīva (鳩摩羅什, c. 409 CE.), 中論 Chinese translation of *Madhyamaka-kārika*; translated into Japanese by SAIGUSA Mitsuyoshi (三枝充悳)(1984): 中論 — 縁起・空・中の思想 (上・中・下), from the Chinese version of Kumārajīva with commentaries of his teacher Vimalaksa (called Piṅgala (青目), Blue eyes).

[45] Candrakirti (c. 650 CE), *Prasannapada*, Its essential chapters translated (in 1979) by M. Sprung in collaboration with T. R. V. Murti, U. S. Vyas, with title *Lucid Exposition of the Middle Way*, new edition, Routledge, 2008.

[46] O. Takeki 奧住 毅 (1988), 中論註釈書の研究 (translation into Japanese of the Prasannapada of Candrakirti, c. 650 CE), 大蔵出版, Okura Publ., Tokyo.

[47] RJE Tsong Khapa (c. 1398 CE), translated by G. N. Samten, J. L. Garfield (2006), *Ocean of Reasoning: A great Commentary on Nāgārjuna's Mulamadhyamaka-Karika*, Oxford University Press, New York.

[48] J. L. Garfield (1995), *The Fundamental Wisdom of the Middle Way*: *Nagarjuna's Mulamadhyamakakarika*, Oxford University Press.

[49] T. Bhikkhu (translation) (2003), *The Great Elephant Footprint Simile* (大象跡喩経, *Mahāhatthipadopama-sutta*), *Majjhima Nikaya* (Pali 中部経典 Collection of Middle-length Discourses) no. 28, http://www.accesstoinsight.org/tipitaka/mn/mn.028.than.html

[50] Tathāgata, Wikipedia, https://en.wikipedia.org/wiki/Tath%C4%81gata

[51] Original Chinese text was discovered at Dun-huang in 1900. English edition, *The Bodhidharma Anthology* (edited by Tan-lin, Huike and disciples of Bodhidharma, about 540 CE), translated by J. L. Broughton, (1999), Univ. of California Press; Japanese edition, 『達摩の語録』(二入四行論) by Y. Seizan 柳田聖山, in Series 禅語録1『達摩語録』, Chikuma shobo 筑摩書房, Tokyo, 1969.

[52] *The Lankavatara Sutra: A Mahayana Text*, translated by D. T. Suzuki [from the original Sanskrit, 1931], first published in 1932 by Routledge & Kegan Paul Lt., London; 1978-version, by Prajna Press, Boulder, Co. USA, 1978.

[53] *The Lankavatara Sutra*, translation and commentary by Red Pine, Counterpoint Press, Berkeley, CA, USA, 2012.

[54] *Jing-De* (era) *Record of Transmission of the Lamp* (景徳伝燈録), compiled by Shi Daoyuan (釈道原, documenting historical records of Chan Buddhists in China), 1004 CE.

[55] *Anthology of the Patriarchal Hall* (祖堂集), compiled by Jìng (靜) and Yún (筠), Chan Masters, 952 CE. This is the oldest existing collection of Chan (Zen) Buddhism, published before the reference [54]; (a) Korean version: This was left as an Appendix to the *Tripitaka Koreana* (高麗大蔵経) in Korea. After being lost for centuries, it was rediscovered in the 20th century at the Haeinsa temple (海印寺) in Korea, in a complete form with all 20 chapters; (b) Japanese edition, 『祖堂集』, translated by Yanagida Seizan 柳田聖山, in the Series 『大乗仏典13』 (Chu-Oh-Koron-sha 中央公論社, Tokyo), 1990; (c) Taiwan edition, 禪宗全書, 史傳伝一, 第七 『祖堂集』, (文殊出版社、台北), 1988.

[56] Jing-jue (淨覚), Chinese Buddhist record 楞伽師資記, Record of Lanka Masters and Disciples, 716 CE.

[57] *Record of Dharma Treasure through the Generations* 歴代法宝記", compiled by an anonymous author, a disciple of 無住 (714–774), excavated from Dunhuang, hence c. 774 or after 774; Japanese edition, 『歴代法宝記』 (頓悟法門) by Yanagida Seizan 柳田聖山, in Series 禅語録3『初期の禅史II』, Chikuma shobo 筑摩書房, Tokyo, 1976.

[58] 六祖壇経, *Platform Sutra of the Sixth Patriarch*, by Fa-hai (法海, a disciple of Hui-neng, 1st ed., c. 720), c. 780 CE. The early version was found in the Dun-huang cave.

[59] *Treasure Forest Record* 宝林伝, by Chiho 智炬. 801. It is the history of Zen *Dharma-Lantern* established in the 801. Originally, there were 10 volumes in total, but the current text lacks vol. 7, vol. 9, and vol. 10.

[60] 伝法正宗記, *Record of Transmission of Orthodox Dharma*, by Qì Sòng (契嵩), 1062 CE.

[61] Buddhist record 伝法宝紀, *Record of Transmission of Dharma Treasures*, 713 CE.

[62] Dao-xuan (道宣), 続高僧伝, *Memoirs of Eminent Monks*, 645 CE.

[63] Figure 10.6 is taken from the image, designed by the monk Gen-sho (玄証 Japan, 12th century), (C0089915 白描先徳図像 — *Tokyo National Museum* A-10_web-search) http://webarchives.tnm.jp/imgsearch/show/C0089915

[64] *Nirvana* Sutra 涅槃経 is one of the major sutras of Mahāyāna Buddhism. It is also called Mahāparinirvāṇa Sūtra. Although the *Nirvana Sutra* mentions some of the well-known episodes in the nirvana-moment of the life of the Buddha, the sūtra uses these narratives merely as a convenient springboard for the expression of standard Mahāyāna ideals. (a) The text of the *Nirvana Sutra* in the original Sanskrit has survived only in a number

of fragments, which were discovered in Central Asia, Afghanistan, and Japan. It does exist in Chinese and Tibetan versions of varying lengths. Fa-xian (335–421 CE), the monk who initially brought back the text to China from India, prepared a brief translation containing six fascicles, but Dharmakṣema's translation slightly later had 40 fascicles. Still later, Hui-guan, Hui-yuan, and others during the Liu Song dynasty integrated and amended the translations of Fa-xian and Dharmakṣema into a single edition of 36 fascicles. That version is called the *Southern Text* of the *Nirvana Sutra*, while Dharmakṣema's version is called the *Northern Text*. The Jataka story is given in the Chapter of *Holy Exercise (Action)* of vol. 13 of the Southern Text. (b) English version, *The Mahayana Mahaparinirvana Sutra*, translated into by Kosho Yamamoto, 1973, from Dharmakshema's Chinese version (Taisho Tripitaka vol. 12, no. 374), edited and revised by Tony Page, 2007.

[65] *Lankavatara* Sutra 楞伽経 is a sutra of Mahāyāna Buddhism. The first extant Chinese translation is the one translated by Guṇabhadra (Section 2.6) in 443 CE, consisting of four fascicles. This edition by Guṇabhadra is said to be the one handed down from Bodhidharma to Huike. The sūtra asserts that this describes the mind of what Buddha told. The *Laṅkāvatāra Sūtra* draws on the concepts and doctrines of Yogācāra and Tathāgatagarbha. (See [39] for Tathāgata.) The sūtra asserts that all the objects of the world, and the names and forms of experience, are merely manifestations of the mind. That is to say, there is nothing that humans experience that is not mediated by mind (*Yogasara*). The *Sūtra* describes the fundamental *storehouse consciousness* (Sanskrit: *ālayavijñāna*), which is the base of the individual's deepest awareness and his/her tie to the cosmic. In this sense, the *storehouse consciousness* is identified with *Tathāgatagarbha*. The sūtra asserts the *vow* of *bodhisattvas*: they do not dwell in the world of continuous fabrications of forming and decaying with the fundamental wisdom, nor they do dwell in nirvana, in order to relieve sentient beings with the great mercy (*mahā karunā*). (See the verses in Section 3.2.2.) The concepts of *all-at-once* dharma (figuratively, like *mirror images*) and *step-by-step* dharma (figuratively, like generation of sea waves, or growth of plants in nature, or learning of engineering techniques) are explained in this sūtra. There are other versions: *Ten-Fascicle Lankavatara*, translated into Chinese by Bodhiruci in 513 CE, and *Seven-Fascicle Lankavatara*, translated by Śikṣānanda in 704 CE. The sūtra recounts a teaching primarily between the Buddha and a bodhisattva named Mahāmati (Great Wisdom) at Laṅkā, the island fortress capital of Rāvaṇa.

[66] Blue Cliff Record (碧巌録). Main core-texts of one hundred koans by Xue-Dou (980–1052), critical remarks and interpretations by Yuan-Wu (1063–1135).

[67] Gateless Gate (無門関, 1228 CE) by Wu-men Hui-kai (Mumon 無門慧開, 1183–1260); Free online library, *The Gateless Gate* (1934) by Mumon, translated by N. Senzaki, P. Reps, A Collection of Zen Koans, https://en.wikisource.org/wiki/The_Gateless_Gate

[68] Koan (公案) refers to a story selected from sutras or historical records, or to a perplexing element of a story, which stimulates awakening or guides to spiritual insight. A collection of koans is used as a textbook for students in training, teachers or monks in Zen monasteries.

[69] Yǒngjiā Xuánjué (永嘉玄覺, 665–713), *Song of Enlightenment* (證道歌 Zhèngdào gē); Chang, Chung-Yuan (1967), "Ch'an Buddhism: Logical and Illogical," *Philosophy East and West*, vol. 17, no. 1/4:37–49, JSTOR 1397043, doi:10.2307/1397043.

[70] Mahakasyapa: wish to practice an austere life, Buddhist Studies: disciples of the Buddha: www.buddhanet.net, *Thirteen Dhutangas: The 13 Ascetic Practices,* http://en.dhammadana.org/sangha/dhutanga.htm

[71] Bi-Guan Brahman 壁観婆羅門: Chinese "*Bi*" is a wall and "*Guan*" is to look or to contemplate. Nominally, Bi-Guan is wall-contemplation. Even now, there is a cave called *Dharma Cave* up on a hill near the Shaolin Temple within which there is a statue of the sitting Bodhidharma.

[72] Sutta Nipata: Discourse Collection in the Theravada Buddhism, consisting of five chapters. All discourses are thought to originate from the life time of Buddha and consist largely of verses. For the English version, *The Discourse Group (Sutta Nipāta)* (2016), translated by Ṭhānissaro Bhikkhu, see https://www.dhammatalks.org/Archive/Writings/SuttaNipata160803.pdf

[73] *Record of Buddhist Monasteries of Luo-yang* (洛陽伽藍記). 547 CE (Probably there, he saw also numerous diligent monks working for translation work.)

[74] Nun-Dharani is said to be a daughter of Emperor Wu [57, 68].

[75] 増演易筋洗髄内功図説、edited by 周氏, 1895: 少林真本、系養生気功著作、全書共十八巻, 少林書局出品。

[76] Shaolin publication (2006), Selected legends and history of Shaolin Temple (少林寺伝説故事選), Shaolin Publishing 少林書局, Zheng-zhou (鄭州), China.

[77] Ye Feng (葉封) *et al.* (1671), *History of Shao-lin Temple,* Shaolin Publication (Zheng-zhou, China, 2007).

[78] Stone monument 689, *Tang Zhong Yue Samana Shi Fa Ru Chan Shi Xing Zhuang* (唐中岳沙門釈法如禅師行状).

[79] 每日頭條 https://kknews.cc/other/85p39pl.html; 「禪宗二祖埋葬地，河北邯鄲元符寺」, 2017-04-04 由 聚邯鄲 發表.

[80] Guang-xiao-si (光孝寺) (2000; 1st ed. 1769; 2nd ed. 1935), History of Guang-xiao-si (光孝寺志), China Publication (中華書局), 杭州富陽古籍印刷廠.

[81] Hui-Neng, https://en.wikipedia.org/wiki/Huineng#Diamond_Sutra

[82] *The Sixth Patriarch's Dharma Jewel Platform Sutra*, 2001, published and translated by Buddhist Text Translation Society (Burlingame, CA 94010-4504) 1st ed. (Hong Kong) 1971; 2nd ed. (USA) 1977; 3rd ed. (USA) 2001.

[83] Fa-xian (法顯) 414, *A Record of Buddhistic Kingdoms* (English translation); *Fa-Xian Story and Song-Yun Story* (translation into Japanese and comments by K. Nagasawa, Toyo-Bunko no. 194), Heibon-sha, Tokyo, 1971.

[84] *Ho-kyo-ki* (宝慶記) by Do-Gen (道元), Records made in Ho-kyo age (宝慶時代) of China, 1227.

[85] Eihei-ji (A collection of photographs), (Eihei-ji 永平寺, Fukui, Japan), 1979;Ryo-an-ji,Wikipedia,https://en.wikipedia.org/wiki/Ry%C5%8Dan-ji

[86] Zazen (座禅) (2007), How to sit, by Daidokai (大道会), Hokyoji 宝慶寺, Ono, Fukui, Japan.

[87] H. Kikuchi, *Visit to see the Gandhara Buddhas exhibited* at the Tokyo National Museum (Ueno), https://www.h-kikuchi.net/entry/2017/01/21/

[88] S. Weinberg (2015), *To Explain the World*: *The Discovery of Modern Science*, Penguin Books.

[89] W. Isaacson (2007), *Einstein: His Life and Universe*, Simon & Schuster.

[90] Tropical cyclogenesis, Wikipedia, https://en.wikipedia.org/wiki/Tropical_cyclogenesis; Wikipedia, Tropical cyclone, https://en.wikipedia.org/wiki/Tropical_cyclone

[91] B. Israel (2010), Hurricane alley heats up with stormy threesome, *Live Science*, https://www.livescience.com/8554-hurricane-alley-heats-stormy-threesome.html

[92] *Life of a Typhoon* (in Japanese) at the HP of Japan Meteorological Agency (JMA), http://www.jma.go.jp/jma/kishou/know/typhoon/1-2.html

[93] *Weather Compact Tidbits* (in Japanese) at the HP of Bio-weather and JMA, https://www.bioweather.net/column/weather/contents/mame092.htm

[94] *The Pallava King from Vietnam,* Live History India, https://www.livehisto-ryindia.com/story/people/the-pallava-king-from-vietnam

[95] A website for a Tamil stone inscription (2019): Rare Tamil stone inscription found in China; sheds light on cultural relations between ancient China and Tamilagam — Simplicity; *Same-website*: https://simplicity.in/coimbatore/english/news/55252/Rare-Tamil-stoneinscription-found-in-China-sheds-light-on-cultural-relations-between-ancient-China-and-Tamilagam

[96] *Roots: Journey to the Origin of Life* (Roots 生命起源への旅), TV program by NHK (Japan Broadcasting Corporation), 10 Feb, 2022, https://www4.nhk.or.jp/P7347/3/; A magnificent story about the first life that appeared on the earth about 4 billion years ago, and how it emerged.

[97] Big Bang, Wikipedia, https://en.wikipedia.org/wiki/Big_Bang

[98] D. Geary (2016), Walking in the Valley of the Buddha: Buddhist Revival and Tourism Development in Bihar, *J Global Buddhism*, **17**, 57–63.

[99] Germination (Google) by Getty Images: https: //www.istockphoto.com

[100] *The Lotus Sutra* (妙法蓮華経), English version, translated from the Kumārajiva's Chinese edition by T. Kubo, A. Yuyama, Numata Center for Buddhist Translation, 2007, https://www.bdk.or.jp/document/dgtl-dl/dBET T0262_LotusSutra_2007.pdf; BDK (Bukkyo Dendo Kyokai - Society for the Promotion of Buddhism). The BDK was recognized as *Public Interest Incorporated Foundation* and the new start began in 2013.

[101] V. N. Fujii (1979), Seal of fearlessness, in *The Most Venerable Nichidatsu Fujii*, https://atlantadojo.tripod.com/gurujipage.html

Index